THE
SCIENCE
OF
GOLF

THE MATH, TECHNOLOGY, AND DATA

Skyhorse Publishing

Skyhorse Publishing books may be purchased in bulk at special discounts for sales promotion, corporate gifts, fund-raising, or educational purposes. Special editions can also be created to specifications. For details, contact the Special Sales Department, Sports Publishing, 307 West 36th Street, 11th Floor, New York, NY 10018 or info@skyhorsepublishing.com.

Skyhorse® and Skyhorse Publishing® are registered trademarks of Skyhorse Publishing, Inc.®, a Delaware corporation.

Visit our website at www.skyhorsepublishing.com.

10 9 8 7 6 5 4 3 2

Library of Congress Cataloging-in-Publication Data is available on file.

Cover design by David Ter-Avanesyan
Front cover illustration by Getty Images

ISBN: 978-1-5107-7185-7
Ebook ISBN: 978-1-5107-7186-4

Printed in the United States of America

CONTENTS

INTRODUCTION

With roots dating back to the mid-fifteenth century, golf has both the blessing and curse of being a sport with much history and evolution. And while the modern playing of the sport is arguably only two hundred years old, the fundamental goal of the game has remained the same as it is written in its official rules today of the United States Golf Association (USGA).

> *Golf is played by striking your ball with a club, and each hole starts from the teeing area and ends when your ball is holed on the putting green.*
> *You should normally play the course as you find it and play your ball as it lies.*

Simple enough, right? Golf's greatest gift—the ability for any player, of any age, or any skill, to be able to compete alongside each other—can often be the sport's greatest source of consternation. How one gets the ball from teeing area to hole is as varied today as it was during its primitive origins on the Scottish coastline half a millennium ago.

The key to success at golf (in whatever measurement of success one aspires to reach) rests in understanding how a variety of forces are at play to make that ball (with a minimum diameter of 1.68 inches) move in the most efficient way via clubs (of varying restriction and construction) around a property of endless obstacles and altering conditions.

To some, this exercise is an art form. Many of the world's golfers arrive at mastering their craft by "feel." They are capable of maximizing their own abilities and the movement of their golf ball through repetition, experimentation, and muscle memory.

Feel players have an organic connection with the game and seek to conquer it not through academic discovery, but rather with a more purist belief of unity with the sport and a link through time to golfers of generations past.

Many others—especially with the technological and analytical revolution in the sport during the twenty-first century—attack the game with as much information as possible. The modern golfer leaves no stone unturned in pursuit of the most efficient path to that hole (of 4.25 inches in diameter).

This book is not about which process is correct; it is simply about the process.

I have been around golf almost all of my life. I owe much of my success—tangibly and immaterially—to the lessons learned from working in the golf business and playing golf both competitively and recreationally. Simple osmosis gave me enough understanding of the game, but the reality is that I was an ignorant amateur when it came to breaking it all down. Amateur, I should add, comes from the Latin word *amator*, which means lover. I love golf. I just didn't understand it the way I thought I did.

It wasn't until 2017, four years after I had started broadcasting golf on the PGA Tour full time, that my curiosity in the sport increased to where I wanted to peel back the many layers of the sport and understand it more.

In my pursuit of understanding how the best golfers in the world reach their goals, break records, and redefine the sport, I stumbled across a seemingly endless amount of ways in which those elite golfers, teachers, and analysts study the game to squeeze every drop of potential out of it and themselves.

This book aims to explore all of those areas, offering a glimpse not just in the "what" of golf, but the "how" as well.

I would be remiss if I didn't acknowledge that there are far more detailed and elaborate scientific studies of many of the principles and subjects discussed in this book. I had once, naively, thought this book would break down the formulas and function of several of the most fascinating studies, but quickly realized that my general knowledge puts me closer to the forward tees than the tips. Many are still referenced, and a few are even authored by the subjects who graciously gave up some of their time to be interviewed for this book.

Rather than chase those topics down a rabbit hole of intense evaluation—and understanding beyond my own humble comprehension—this book aims to give the reader a broad understanding of all of the scientific factors in play in the sport, and view them through the lens of some of the world's best players and minds. Like golf itself, this exercise could be endless, but we want to keep up the pace of play.

Because what makes golf beautiful is also what makes it so different from reader to reader. Hopefully, you can get a better sense of the "what" and the "how," and it leads to a better "why" for your game, and your scores!

CONTACT

Zero-point-five milliseconds. Written in more understandable terms: 0.0005 seconds. The golf ball interacts with the face of a golf club for approximately 0.0005 seconds. That is eight hundred times faster than the time it takes an average human to blink, and just sixteen times slower than the time it takes lightning to strike.

This incredibly short moment in time also happens to be the most important moment of the sport. For what happens in this 0.0005 seconds begins a domino effect of cause-and-effect reactions. It determines where the ball goes, where it lands, how it lands, and, most importantly, what the next shot will be. Ultimately, over the course of many 0.0005-second impacts, a golfer arrives at a score, represented by the total number of those impacts.

For example, in a four-hour round where a player shoots 80, the most important actions impacting that score of 80 take place for a combined time of less than half of a tenth of a second. Blink, and you miss the entire round of golf.

While golf has had numerous historical advances in its understanding and execution, none may have had more impact on how experts study (and teach) the sport than the introduction of the launch monitor in the twenty-first century. Prior to the technology's arrival in the sport, understanding contact of the golf club with the golf ball was empirically gathered by studying the result of that contact, not necessarily the cause. What did the ball do after contact? Not, what did the club do to the ball?

Radar and launch monitors moved the microscope away from the effect and showed us everything that was happening at the moment of causation . . . contact.

Popularized largely by brands like TrackMan, FlightScope, Foresight, and others, harnessing existing technology to study the interaction of the golf club with the golf ball, plus the subsequent flight of the ball after impact, has turned that causation discussion of the game in an opposite direction.

Prior to this century, golf was a study of how a golfer's movement led to the result of a shot. Stance, grip, takeaway, backswing, top position, downswing, and follow-through were just a few of the many elements that were perfectly choreographed in order to make ideal contact with the golf ball. How you swung the club determined what the ball did.

Today, while that statement is true, the growth of understanding of launch conditions and what the ball does has shown that there are many ways the swing can be choreographed to achieve similar or identical contact conditions.

"Science has played a larger role in golf instruction in recent years, and it is going further and further," says Mark Immelman, golf instructor and analyst for CBS Sports. "When I began instruction, the science of it was whatever video camera you are using, and you got as much video as you could. You'd put stuff side by side and you just compared positions. Then there was the advent of the launch monitor."

Launch monitor technology was already used in the non-sports world, but taking that tech and applying it to sports with a projectile component made too much sense. If the technology already existed to track a missile traveling at rapid speeds in the air, why not a golf ball? That was the case for Henri Johnson, who founded FlightScope in the late 1980s for military purposes and eventually leveraged that technology to advance scientific discovery in the sports world.

"It keeps you focused on the outcome," says Luke Kerr-Dineen, who covers golf technology, the swing and improvement. "If there is a negative to geeking out too much into a video camera, it's that you start looking at your own golf swing and not what the ball is doing. Whereas with TrackMan, it's all about what the ball is doing. It's all about making sure you are understanding how that ball is moving and why, and I think that those are the first principles that constantly need to be highlighted and focused on by golfers."

The result from developments in the launch monitor sector is a variety of radar-powered technologies that can track a seemingly endless amount of data points.

At contact, that includes data points like:

- **Club head speed**—The speed of the club at the moment right before impact (and in some cases, the speed immediately after impact to help calculate energy transfer and loss).
- **Ball speed**—The speed of the ball as it leaves contact with the club.
- **Smash factor**—A simple calculation of dividing ball speed by club speed. This ratio is then used, depending on the club used, to determine the quality of strike by means of the efficiency with which energy was transferred from the club to the ball. This has become a very important measurement in how golf equipment is regulated, as the USGA has capped the smash factor ratio for approved driver technology at 1.5 in club testing.
- **Ball spin rate**—The backspin put on the ball at impact, measured in revolutions per minute (rpm). This measurement can vary depending on type of desired shot and outcome, but typically is desired to be low for drivers and increasing for more lofted clubs through the bag. (Golf ball technology has had a major role in this, to be discussed in chapter 6).
- **Sidespin rate**—How much spin, in rpm, is generated on the ball in a horizontal direction, versus the vertical direction of the backspin. This spin determines how far the ball will move either right or left of the intended target line. It is a

measurement for how much slice (left-to-right movement for a right-handed golfer) or draw (right-to-left movement) there will be on the shot.

- **Spin axis**—Like smash factor, this is a combination of two data points, backspin and sidespin, creating a positive or negative reading depending on whether the ball is spinning to the left or to the right. The smaller the number (the closer it is to 0), the straighter that golf ball is traveling. To understand that visual, take this example from TrackMan: "The spin axis can be associated to the wings of an airplane. If the wings of an airplane are parallel to the ground, this would represent a zero-spin axis and the plane would fly straight. If the wings were banked/tilted to the left (right wing higher than left wing), this would represent a negative spin axis and the plane would bank/curve to the left. And the opposite holds true if the wings are banked/tilted to the right.

  In general, a spin axis between -2 and 2 can be considered a straight shot. Under normal conditions, it would be difficult to see curvature on a shot with a spin axis between -2 and 2. The higher the number of the spin axis, the more curvature should be visible."

- **Attack angle**—This angle is in relation to the ground and measured as the club is arriving at impact with the ball. As will be discussed further, that angle has grown to be, generally, upward for swings with a driver, creating a positive attack angle. Negative attack angles measure a downward strike, generally making contact with the ground, the typical result with irons and wedges.

- **Launch angle**—Simply put, the angle of the ball's takeoff from contact with the club in relation to the ground. Not as complicated as some other measurements, but arguably one of the most important data points for golfers of all skill levels as this measurement, coupled with spin rate, is most important for maximizing the efficiency of swing, equipment, and contact.

- **Face angle**—The most important measurement when it comes to understanding the starting direction of the golf ball. This number is positive when the face of the club is pointing to the right of the target (described as "open" for right-handed golfers), and negative when the clubface is pointing to the left ("closed"). A face angle of zero would be the easiest path to a straight shot, but straight shots aren't always demanded or desired, so manipulation of the face angle is also one of the most important measurements to creating various shot types and paths.

- **Face loft**—This measures the actual loft of the club at impact. Each club has a static loft, a loft of the face of the club that is measured in relation to the ground when the club is at rest. Launch monitor technology can measure the actual loft of the club at impact, which can change from its static loft due to a number of factors, such as attack angle or the shaft in the club. This data point has been developed into an even more detailed measurement called "dynamic loft," which adds in angle of attack and even where the ball strikes the clubface to determine the loft angle of the ball. Not to be confused with launch angle, understanding how dynamic loft leads to launch angle and shot result is a major part of the process.

- **Club path**—This measures the direction the club is moving at impact in relation to the target line. A positive value represents a path wherein the club is moving that is to the right of the target at impact (commonly referred to as an in-to-out movement for a right-handed golfer). A negative value would be a path moving to the left of target (out-to-in).

- **Face to path**—A measure of how much the club's angle is on the same angle as its path at impact. This relationship between two angles is critical to the understanding of ball flight.

From there, ballistic radar (or calculations, depending on the sophistication of the product) can give another set of data points to measure what the ball is doing after contact. Some of those data points include apex height of the ball's trajectory, shot flight shape, carry distance, roll-out distance, and many more. While those are influenced by conditions, most notably wind, it still comes back to that instantaneous moment of contact. Optimize that split second, and golf gets easier.

"We've got so much data now at our fingertips that it's really not guessing," said Alex Trujillo, who works with elite golfers on the PGA Tour for FlightScope. The addition of this technology provides instant feedback and understanding for every golf swing about what the golf ball is doing. It has accelerated the learning process and helps make the best in the world even better.

"I like to call [the radar] the MRI. It lets us figure out in detail what is happening in these circumstances," Trujillo adds. "To tell you the truth, even a guy like Bryson [DeChambeau] or Bubba [Watson] . . . they're not always, day in and day out, working on face and path. They have their swing. They've grooved it. If there's a minor tweak that they needed, they're going to look at that on radar, but they're really just looking at what the conditions are today. What are the temperatures? How far am I hitting this golf ball? What's my spin rate? And what's my launch angle? They're all looking at that window everybody talks about. If you've watched them long enough, and you stand behind them, when they hit a ball, when they look up, they know, okay, that ball's in that window."

At the highest levels of professional golf, that is the scene on every driving range. A world-class player, a team of professionals that understand both the golf swing and, now, the data, and a high-tech piece of technology sitting on the turf next to their golf balls, getting ready to measure every detail of each 0.0005-second strike.

The result of that technological revolution of the twenty-first century has had trickle-down effects across the sport. Not only are the best players in the world better because contact (and club technology) has been optimized more than ever before, it has also simplified the realities of physics as it relates to the golf swing and how it is taught to all skill levels.

"Even if you're old school and you're shouting at the new-school kids that have got nothing but launch monitors, you would be a fool not to use this stuff now," Immelman said. "Because it eliminates conjecture."

Lee Trevino is considered to be one of the purest strikers of the golf ball in the history of the game. That's an opinion shared by many, including the greatest iron player—anecdotally and (modern) statistically speaking—of all time.

"He finds the middle of the [club] face each and every time," Tiger Woods said of Trevino during a press conference in advance of the 2021 PNC Championship. "No one has controlled that golf ball as well as he has."

Big praise for one of the best self-taught golfers of all time. Trevino famously grew up in a blue-collar household and learned the game of golf sneaking away from his work at a nearby country club to bash balls over and over again from hardpan dirt.

Those conditions left little room for error and provided Trevino with immediate feedback on how the golf ball interacted with the face of his clubs. He didn't have fancy equipment, expensive instruction, or even normal access to golf facilities, let alone the modern technology of a launch monitor. He simply had his repetition and a natural-born sense of feel that built a lifetime of understanding how to create perfect contact with the golf ball.

In giving a lesson throughout an interview with Kerr-Dineen for Golf.com from the practice tee of the Berenberg Invitational in October 2021, Trevino broke down his visualization and execution of shaping a golf shot.

> *My thought when I hit a golf ball is I am always trying to hit it at 7 o'clock. This golf ball is a clock. You've got 6 o'clock (points to the middle of the back of the golf ball, the Y axis point), 5 o'clock (points a few degrees to the right) and 7 o'clock (points a few degrees to the left). I am trying to hit the ball on the inside because your hands are going to be past the ball when you hit it. If you're going to cut the ball, you hold [onto your hands]. If you are going to draw the ball, you rotate them. But you are hitting it at 7 o'clock.*

What Trevino just described is his process as a player to make sure he is repeatedly swinging the club on a certain path. He guaranteed his striking point at his 7 o'clock position on the ball almost every time, programming the muscles in his body to repeat the same club path through (likely) millions of swings in his lifetime. It was something he could control to assure a solid strike.

But it was the subsequent manipulation of his clubface angle that would make the ball do what he wanted. Most of his lesson was about where to hit the ball, but the magic that was somewhat hidden was in those two lines about whether he held on or rotated his hands.

What does science tell us now about this process? It tells us that Trevino is right, if you were paying attention to those hands.

"The launch monitors have proven that a draw [right-to-left ball flight for a right-handed golfer] happens because the alignment of the face is pointing to the left of the path of the swing," Immelman says. "And the path is how the club is approaching into impact. So, if the face is left of the path, the ball's going to turn to the left. It's just the rule of it now. And if the face is to the right of the path, the ball's going to turn to the right . . . You can make a swing that looks like it is going to draw, but it fades because of how the face is in relationship to the path that it's swinging on."

Having real data from launch monitors changed the old laws of how swing manipulation was thought to influence the direction and shot shape of the ball. There are several detailed, tested, and in-depth pieces that breakdown this study. One of them, from a review paper of Jeff Mann, highlights how a majority of the ball's initial flight direction is determined by the clubface orientation. This differed from predictions prior to the introduction of radar data that the path of the swing was the key indicator to how a ball would take off and travel in the air. In the end, Mann's conclusion:

The fundamental principle of the "new" ball flight laws is that the ball starts off approximately 85-percent in the direction of the clubface orientation if there is a divergent angle between the clubface orientation and the club head path, and the ball will curve away from the club head path. If the clubface is closed to the club head path, then the ball will curve to the left (draw); and if the clubface is open to the club head path, then the ball will curve to the right (slice).

But what about the contact itself? Launch monitors proved how the directional path of the club and the angle of the face impacted the linear path of the ball, but what about the three-dimensional space (launch angle) and spin?

The collection of millions of data points from golfers of all skill levels has helped to form a more concrete understanding of ideal launch conditions. While this is heavily influenced by a golfer's skill and speed, plus equipment (which is explored in later chapters), there is a literal and figurative sweet spot from which ideal contact can be judged and determined.

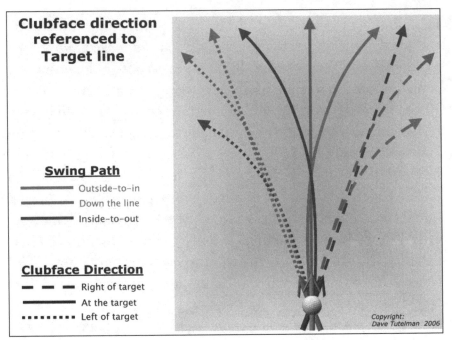

Clubface direction referenced to Target line

Swing Path

———— Outside-to-in
———— Down the line
———— Inside-to-out

Clubface Direction

— — — Right of target
———— At the target
•••••••• Left of target

Copyright:
Dave Tutelman 2006

The lessons learned from launch monitors provided absolute data on how the clubface and swing path relationship influence the flight of the golf ball, as illustrated in this chart showing those shapes in relation to the target. *Original image by Dave Tutelman Via www.tutelman.com*

"I don't think the average player really tries to dive into how efficient their strikes are," Trujillo adds. "Let's forget about the information for a moment. If we go on the range and we start hitting 7 irons, can I, ten balls in a row, hit the same spot every time on the face? I think that if people start to focus on that while they go out to practice, they're going to start becoming better players."

The first goal of hitting that sweet spot is to get as much energy transferred from the club to the ball. Newton's First Law of Motion is quite simple. An object will not move unless a force acts on it. A golfer's mission is to enact as much force on the golf ball to make it move. The body (explored heavily in chapter 2) creates kinetic energy generated in the swing of the club, but as the USGA points out in its breakdown of forces "not all of the club's kinetic energy transfers to the ball. Some of it stays in the club, which keeps moving forward, now more slowly. Some transforms into thermal energy (heat and friction) or sound energy (the little thwack you hear). Some gets briefly stored in the ball as elastic potential energy."

Getting as much energy from club to ball is, essentially, the most important aspect of contact.

The second goal of contact with the sweet spot is to produce the optimal amount of spin to maximize the chances of creating a desired flight of the golf ball.

"At its essence, golf is a game of spin," Immelman says.

Some players want the ball to launch lower with more spin, while others would prefer a higher, "softer," shot with less spin. This can be personal preference or, in most cases, dependent on what the shot demands.

"The part about spin specifically really came into focus when I got on the PGA Tour," says Roberto Castro, who played for nearly a decade on the PGA Tour and graduated with an engineering degree from Georgia Tech. "It took me a long time to improve at [managing spin]. It comes out of contact conditions, how you're striking the ball. A launch monitor can help, but sometimes you don't need a launch monitor to stand on a driving range in the wind and watch what your ball does. You can tell when you get steep (attack angle) and you hit down on it harder, it spins more at the ball. There's more variation in where it goes."

Like direction can be manipulated with the angle of the clubface at impact, the attack angle creates the same manipulation when it comes to spin. This manipulation is also greatly enhanced by its relationship with the ball and equipment, a discussion for later when it comes to the technological understanding and evolution of equipment.

"In the final analysis, it's about applying the clubface to a ball in the desired manner in order to get the ball to spin in a desired fashion," Immelman concludes. "But it's not just backspin. That's sidespin as well. What [the best players] are trying to do is to put the club, or move the club, through the ball in such a fashion to spin the ball in a desired manner. That's really what it is."

Measuring contact with the sweet spot used to be done exclusively with impact tape, which would register the contact point on a golf club with a shot and see where that impact was taking place. If that carbon impression was in the middle of the club, it was thought that contact was optimized and the club was set up properly. While mostly true, it was not a thorough or valid process in terms of measuring energy transfer.

That's where smash factor, that ratio of club speed to ball speed, has become universally accepted as the true measure of good contact.

"Smash factor and ball speed, I kind of interchange them," says Nick Sherburne, the founder of Club Champion, one of the largest club-fitting companies in the world, which uses TrackMan to measure the swings of those who come to get fit for the most optimized equipment. "I want that maximum ball speed, but smash factor helps me know if I'm there. When I am fitting that 6 iron, I want to see that thing at 1.38 [smash factor] or above. And then when I'm fitting a driver, I really want to see it at 1.48 or above. And when I have those smash factors, I know I have this person in the money zone."

The money zone is (nearly) perfect contact within the parameters of current equipment regulations. And the beauty of that measurement is that an 80-year-old golfer who swings the club at 70 mph can have an identical (or better) smash factor than a 25-year-old professional golfer swinging the club at 120 mph. It all comes back to contact.

What has happened through the collection of millions of data points is that a general understanding of what is optimal for golfers of all skill levels has been developed. In 2021, the average PGA Tour smash factor with a driver was 1.49 with an average club speed of 114.4 mph. In a 2019 TrackMan study looking at the average male amateur (a handicap of 14 or 15), the average smash factor was 1.42 with an average club speed of 93.4 mph. It shouldn't be a surprise that contact by the professionals is more optimal than amateurs.

Missing the sweet spot happens more often than not and it isn't just a lack of optimized energy transfer when that occurs.

"If the club is impacting the ball directly on the sweet spot, then all of those forces are going in the direction of the swing and the direction of the loft," says Chris Voshall, a golf club engineer for equipment company Mizuno. "If that ball is impacted on the toe side of the sweet spot, then you're applying a torque in the toe direction, so that force is going to cause a gear effect, which is going to cause the ball to launch more to the right (for a right-handed player) and have a bit of left spin on it because of that gear effect.

"That gear effect happens in every direction, so if you hit the ball lower, below the sweet spot, that club is going to torque in the downward direction, like de-lofting, and that is going to impart a lower launch angle and added spin to the ball. The converse of that high impact is high launch, low spin [a flier lie, for those familiar with the term]. Heel impact is going to be less launch, more spin to the right."

Five hundred microseconds at the perfect collision between two center of mass points on a club and a golf ball and, voilá, you have mastered the game. Or, sadly in most cases, you have not.

Back to Trevino and Woods, who were practicing on the range next to each other at that same PNC Championship in December 2021. A two-minute interaction between the superstars was captured by the PGA Tour's media team, where Trevino and Woods got into a conversation about hitting 50-yard wedge shots where the primary objective is to eliminate trouble to the right.

Trevino walks Woods through his process, involving having his hands pressed forward before he takes the club back. The intended result here, with his 7 o'clock contact point (club path), is that it makes it nearly impossible for his clubface to be open enough at impact to create a shot that will start the ball too far right of the target line and/or spin the ball with a spin axis that would slice to the right.

Woods hops in and offers his take on his own swing. Because his hands start slightly more to the inside, his club path is different than that of Trevino's, so he shows a need to alter his club path ever so slightly higher so that contact will be less likely to produce a combination with his face angle that makes a right miss possible.

Two different swings. Two different approaches. One manipulates face angle. The other manipulates club path. Two of the best ballstrikers of all time arriving at the same point via completely different methods.

"Most [elite] golfers you'll see path tendencies, swing directional tendencies," Immelman says. "And then it's just their job to organize the face inside of that marker, and to have the ball start on the appropriate line and then hopefully curve to the target or stay on the intended line."

The liberating part of the Woods-Trevino conversation to the average golfer, when combined with the data discovery that contact is the true measure of success in the golf swing, has led to new era of discovery and improvement.

The understanding of all of the factors at play at the moment of impact has reverse-engineered how teachers and analysts look at the golf swing. Prior to seeing the radar data and getting some absolutes, it was mistakenly thought that certain movements in the swing, or setup positions, directly correlated to how a golf shot was hit. The means justified the ends.

Today, thanks to seeing the ball and club interact in all of these data points, the ends now justify the means. That's not to say certain habits or setups don't typically lead to certain paths, angles, or contact, but eliminating that absolute has allowed for a lot of flexibility in the way the golf swing is taught.

"If you look at the PGA Tour, I can walk you down the range and show you fifteen different types of back swings, if fifteen guys are out there," Immelman says. "It's just that science has allowed us to prove that if my left foot pops off the ground, a la Patrick Reed, that's not a bad deal, *if* that allows me to do what I've got to do to get the club on the ball in the desired fashion.

"I'm looking for repeatable relationships between swing path, clubface alignment, and angle of attack. The speed is the spice of the meal, if you will."

But how does a swing become repeatable? What needs to happen within infinite unique swings to get the club on the ball in a desired fashion? And, if speed is the spice, the thing that makes it all work even better, how do we maximize that?

You learn everything about what the body can, and can't, do.

BODY IN MOTION

Better golfers just move themselves around more.

—Nick Clearwater

Jeehae Lee won her second Ivy League championship at Yale in 2006 and was on the LPGA Tour just three years later. She spent multiple seasons playing as a professional around the world before transitioning into the business sector. Throughout that entire golfing journey, she was always tinkering.

Lee wanted to shallow out her downswing. She wanted what every golfer, at any level, wanted. She desired to eliminate her mistakes and make a swing that provided her with what she thought was more ideal contact conditions. She wanted her club path to move from outside-in to an inside-out delivery. She wanted to get to Lee Trevino's spot on the clock.

"I kept trying to shallow out my downswing, but did it the incorrect way, because I didn't really understand all of the different components involved with 'shallowing out my swing plane,'" Lee recalls. "I kept getting more and more side bendy on the way down, because that's one way to fake a shallow swing, right? If you just make your right side lower on the way down naturally your hands and clubs are going to fall below the plane [and come inside]."

The consequence to what she was doing to her swing was the stress it would put on her manipulation of the clubface. Golfers like to use the term *flipping it* to describe the movement of turning the clubface aggressively—often more than naturally—in order to make sure the face angle at impact isn't so open that the ball travels (for a right-handed golfer) right, way right. Flipping it saves the shot, but carries risk due to its need for impeccable timing. Flip it late, it's heading right, as we know from the relationship between clubface and path. Flip it too quickly, and the ball will hook.

"All sorts of problems," Lee remembers. "I hated seeing my swing from face on because it looked hideous. I was doing some terrible things to my swing. What I now understand is that to shallow out the swing, you have to have this move back towards the target earlier on to give yourself the stage for your arms and hands to come down."

Lee had to understand how her body worked in motion to get to the position she wanted with that club path. Just like optimizing the contact of the clubface with the golf ball, there was an optimization that needed to happen in her body's movement. For Lee, it was making sure her upper body stayed on top of her lower body, seeing the relationship in where her center of mass was and how it needed to follow the lead of her lower body. It kept her movements in sync, allowed her to stay aggressive on the way down, and, thankfully, never have to flip it to create the proper clubface angle at impact.

Today, Lee is the founder and CEO of Sportsbox AI, a company that has developed an app that provides three-dimensional, biomechanical breakdowns of individual swings simply by capturing a video of the golfer on a smartphone. It is a simpler option that gives all golfers the opportunity to see their swing broken down into a measured avatar of movement. It is the next technological advancement in what has become a quickly evolving science in the sport. The swing can be unique, but the way the body moves within that swing can be perfected.

"When I grade a swing, I look for three criteria," Mark Immelman says. "The first is a solid and square strike on the golf ball. The second is the ability, as much as possible, to repeat under pressure. And then the third is to leverage one's self as much as possible; to create and generate as much club head speed as possible with minimum, repeatable effort."

The last two criteria in Immelman's grading process is where the body comes in. No two swings can be perfectly identical when no two bodies are perfectly identical. Synching up the movement of the swing is, arguably, the most important thing in unlocking an optimized strike. Knowing how the movement works in the first place has become the key to that lock.

Similar to measuring the moment of impact with the golf ball, advanced technology of the past quarter century is at the forefront of understanding the movement of the body (and golf club) in the swing. The game has come a long way from still images or high-shutter speed photographic overlays to see how the body's movement influences the club and swing.

"In the early days we didn't even really know what we were measuring," says Tony Morgan, who spent sixteen years in the biomechanical space as the managing partner for K-Motion, an industry leader in wireless, human motion learning for a variety of sports. "We all say, 'hey, my timing is off.' We could finally put some [motion] sensors on the body, and then say, 'hey, here's exactly what your timing is or here's what your sequence looks like.'

"A really cool development has been actually comparing people to what they do when they're playing their best. And it's very different than you asking what I'm doing wrong. Say you just hit a great 7 iron 170 yards with a high draw. You didn't do it every time, but you did do it once, so you are capable of it. How do we get you to do that more often? That's one of the biggest advantages of the technology now and how people are using it."

Instead of asking golfers to fit their proverbial round swing (what they do best) into a square hole (what a top golfer does best), understanding the strengths and limitations of one's body has been a key to individual optimization. While there are some absolutes that every good golf swing has, learning what each individual swing has to offer is the modern evolution of instruction.

"There's a huge problem that we've had in golf with just learning, in general," says Nick Clearwater, vice president of instruction at

GOLFTEC, which has been using computers and technology to help golfers improve since 1995. "The way we've always learned is that someone who was good at the game would write down an idea of what they thought that someone did. There was no measurement behind it. There was really no scientific basis for any of it, and then that's carried down from golfer to golfer for as long as golf instructors have been around."

One of the pioneers in unlocking the secrets to a body's motion is Dr. Phil Cheetham. A former gymnast, Cheetham has spent most of his professional life studying the body in motion and making athletes better. His fascination with flipping and spinning faster in aerial sports created a natural segue to golf, where rotational speed is at the core (pardon the pun) of a good golf swing.

"The ball doesn't care about your biomechanics," Cheetham says. "All the ball cares about is if the clubface is hitting it in the correct manner. Yes, there are multiple ways of swinging the golf club. You can get away with all kinds of stuff as long as the club head hits the ball correctly, but there is an efficient way to do it. And that that's the kinematic sequence, being optimum."

Cheetham's work on the kinematic sequence is extensive, with detailed research papers, dissertations, presentations, and resource documents readily available for consumption online. They are the product of decades worth of data gathering, which has only gotten better as the science has improved to track movement and allow golfers of all skill levels to learn that they are capable of great improvement within the confines or limitations of their bodies.

Cheetham started in the 1980s with an image-based system, using reflective markers that were affixed to various parts of the body (and club) to capture motion with a camera that could be converted into a three-dimensional computer animation.

From there, wanting to collect even stronger data points and not periodically lose the finicky markers in an image-based system, Cheetham and his peers worked with the electromagnetic

Advanced Motion Measurement (AMM) system, which provided immediate feedback, including measurements of directional and rotational motion with 240 samples per second. While it was cumbersome and involved wires coming off of the sensors themselves, the 3-D avatar looking back at Cheetham on the computer screen now could move ("dance," as he liked to call it) in real time. The kinematic sequence was mapped. From Cheetham's own *Basic Biomechanics for Golf*, here is the scientific breakdown of what the body is doing:

> *In the golf swing our muscles convert stored elastic and chemical energy into muscular force which allows us to generate a well-timed golf swing. An efficient swing requires us to convert or transfer these various types of energy into motion. An efficient swing requires that each muscle fire and generate force with precise timing to generate and transfer energy to each subsequent body segment in the chain. The energy starts from the ground up. The stronger, larger muscles of the legs and core accelerate themselves and the segments above them by pushing on the ground, then in sequence the smaller, fast muscles of the shoulders, arms and wrists fire next to propel the club at maximum speed into the ball. This process has several descriptive names. It is known as proximal-to-distal sequencing, or the kinetic link, or the kinematic sequence . . . It is a basic principle of human motion when the goal is to speed up a distal segment such as the foot, hand or club.*
>
> *In golf, where the need is to create maximal speed of the club, we find through motion analysis techniques, strong evidence of the kinematic sequence . . . During the downswing, all body segments must accelerate and decelerate in the correct sequence with precise and specific timing so that the club arrives at impact accurately and with maximal speed.*

Something that needs to be discussed a little bit more here is the concept of speed. To this point, it has been referred to, or assumed, as important without answering why. At its most basic level, speed is needed to generate enough force to provide the energy transfer (discussed in chapter 1) which makes the golf ball travel. It is obviously implied that the more speed you can create, the farther the ball will go, increasing the probability of getting closer to the hole and finishing with as few strokes as possible.

But modern statistics (to be discussed later) have only increased validity in the value of distance, making the pursuit of speed—even at the cost of some control and accuracy—the modern focus of the game. Speed is also necessary to create the aerodynamic lift in the golf ball. More speed creates more potential for variety. That speed is relevant to the size and strength of the golfer, but no matter the skill level, more speed makes it easier to create dynamic golf shots. And that relevant speed can be increased with a better understanding of the kinematic sequence. Because of that truth, the work done by Cheetham and others to maximize the body's ability to generate the most speed in the most efficient manner has become increasingly more important to how the game is taught and developed.

Before uncovering the four points of the kinematic sequence, it is important to know where the energy is coming from.

Much like Newton's laws of motion guide the contact of the golf club with the golf ball, the human body is a static object that needs force (like the ball) to move it. The only thing a golfer's swing comes in contact with is the ground below his feet. This is ground reaction forces. Again, from Cheetham's *Basic Biomechanics for Golf*:

The golf swing is mostly rotational, so in addition to the side-to-side forces we have to generate rotational forces on the ground. We do this by pushing forward with one foot and backward with the other causing a "force couple" and creating a torque that causes us to turn.

Torque is the most powerful of the three forces at play in generating speed and power in the golf swing. The more torque generated throughout the swing the more energy can be transferred into the double pendulum effect that the arms and club are providing in unison, plus the centripetal force of the club at the outer edge of the circle of rotation. It's a lot of moving parts, driven by torque, which is fueled by a synchronized dance the feet must have with the ground.

Beyond the mapping of the movement of the body, the increased study and use of force plates has given golfers more information on what drives effective body movement and maximizes their swing. Force plates measure where weight is distributed in both feet and how that force being applied into the ground creates the reactionary force from the ground to turn the golfer.

For a right-handed golfer, this meant that torque rotation necessitates force applied at the front of the left foot and at the back of right foot. This winds the body, pulling the ground reaction forces up through the body and creating torque points in the pelvis, chest/core region, and shoulders. Flexibility will determine the degrees in rotation, but it's a windup every golfer experiences, with a majority of the weight—a golfer's center of mass—being bore on the back leg/foot. Simply understanding where the weight is distributed through the feet can give golfers an edge in maximizing energy forces.

"I think a myth that's been perpetuated throughout golf is that if you're inflexible you can't perform the most basic functions of the swing," Clearwater says. "I've never met someone who couldn't turn their shoulders ninety degrees, couldn't turn their hips forty-five degrees. Both of those are about the average you're going to see each week on the PGA Tour, and I've never met anyone who couldn't tilt their shoulders to the left or to the right adequately enough."

Certainly, flexibility is a massive component to the ease with which golfers can get into certain positions of the swing. Flexibility and mobility is a major part of generating swing speed and helping to avoid injury. And while being inflexible is not ideal to maximizing gains and longevity, it's also not a dead end to arriving at an optimal swing for an individual.

Timing up the movements of the swing perfectly on the way to impact becomes the biggest challenge, but also a universal solution any golfer can discover to open the swing's fullest potential. In order to get all of that energy back down towards the ground and transferred from body to club to ball—and from back to front leg/foot—another quicker choreographed dance must now take place.

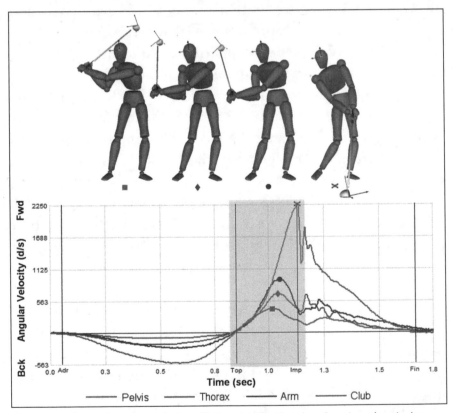

Cheetham's work captured the efficient golf swing by showing the timing and acceleration of various points of the body and how that sequence fired throughout the entire swing *Via Basic Biomechanics for Golf*

That brings us back to the discovery of golf's kinematic sequence and the data which modeled Cheetham's most efficient way to swing a golf club. After loading up the body with power, what is the order the body follows on the way down to contact?

The pelvis drove the downswing first, and it came with a trigger.

"One of the things we've seen is that sway comes before turn at the beginning of the downswing," Cheetham says. "One hundred percent of top pros do that, sway back to where they were at address before impact."

There is no simultaneous transition, as evidenced by the acceleration chart mapping the entire kinematic sequence. In fact, as Cheetham's research found, the downswing followed the takeaway in reverse order. On the way back, the club leads, followed by the arms, twisting the shoulders to drive the midsection and finally the hips/pelvis.

On the way down, the transition started with the sway, followed by the rotational turn of the pelvis, then thorax, then arms, and finally club. This chain reaction fires muscles from one group to the next, allowing speed to increase throughout. Like the smash factor of a club with a ball, putting these motions in the correct order created the most efficient energy transfer throughout the body to the club.

Again, the ground plays a huge role in this, even more so than in the takeaway, when it comes to speed. Channeling ground forces into the downswing has been one of the most important discoveries of motion when it relates to the modern swing.

"I never realized that what you're trying to do is essentially push into the ground as hard as you can, and that creates the force that you need," says five-time PGA Tour winner Ben Crane. "I've been looking at foot motor patterns and what guys do. I've always been kind of an upper body swinger, but the best players in the world who hit it the farthest use the ground so effectively, like Rory McIlroy or Dustin Johnson. They just use their feet so well. It's just a full athletic move."

The push that Crane described is the opposite move of the takeaway. On the downswing, that means the front of the foot of the back leg now drives the uncoiling of the rotation created on the backswing. The more force that can be pushed into the ground with that back leg, the more that can be transferred into the downswing through all of the kinetic movements.

The understanding of ground force energy is now a universal truth in the modern sport, as Crane and his fellow veterans came to understand with younger, faster players becoming the norm in the sport, leveraging the forces of the ground to propel all of their motion in a faster way.

What isn't universally understood (at least at the writing of this book) is the speed with which the takeaway leads to the transition in the downswing. Some studies suggested that a faster backswing with a quick transition to the downswing was able to transfer more energy and speed to the shot. There was more linear (connective) energy without a full pause. It came with the added risk of timing needing to be even more perfect to transition seamlessly into the kinematic sequence, but any more energy that can be added the better, right? A slower, more deliberate takeaway is the more traditional approach, and likely offers timing benefits in terms of the sequence on the way down. Even one of the most aggressive pursuers of speed and power has wavered a bit on which is better.

"That's why when people swing it slower and softer, they go faster . . . because there is no tension," U.S. Open champion and the PGA Tour's longest hitter, Bryson DeChambeau, said in a late 2021 podcast interview with Mark Immelman. "The tension is what absolutely ruins the stretch-shortening cycle. It's like a rubber band, pulling it back and letting it go. When people swing smooth and take it a little longer back and take more time at the top, that gives them the ability to accelerate the club through the stretch-shortening cycle a lot more efficiently and they can hit it farther."

While the transfer of speed may still be in debate, the necessity of speed on the way down can not be denied, and it has become the most important force in every aspect of the sport. No matter the swing speed to start, gaining more speed will only help.

The discovery of the proper kinematic sequence allowed golf teachers to personalize results to each player. While every move in the kinematic sequence may not look the same, the timing of those moving parts could always be choreographed in the same fashion. That knowledge fixes swings without rebuilding them. Timing, as they say, was everything.

For Lee, a second discovery in her own journey to finding a better swing was a simple physical marker of pressure in her left toe. That came from understanding the need to sway and turn the pelvis first. Her pelvis was rotating greatly in her backswing, the product of her own above-average flexibility, but in her early attempts to shallow the downswing, her pelvis was out of sequence in her personal kinematic sequence. She found that putting pressure into her left toe created a swing where the weight transferred her pelvis back to a neutral number, allowing the final three movements in her sequence to fall perfectly in line. By seeing her body move in a 3-D space, she was able to develop a unique, personalized trigger that allowed her body to dance the right way. Lee wasn't thinking about all four rotational elements of her swing every time she took the club back. She used the knowledge of her body to come up with a simple trigger that has now led to a more efficient move.

"If you went and you looked at the speeds, a lot of times you'd see the lower body would be much slower," Morgan adds, reflecting on data collected from thousands of kinematic sequences for golfers of all skill levels. "You'd see amateur players that have very low pelvis speeds. They'd be very dominant with their upper body, very dominant with their arms in the downswing and maybe resulting in an outward [club] path of the hands. They hit pulls and slices.

We would see with the kinematic sequence, or with the speeds, that if we could get that lower body moving better, and it matched up with what the club was doing, we could really help amateur golfers get better."

Like with the knowledge gathered on the moment of impact, the reverse engineering of the body's movements is uniquely tailored, but also rooted in some of the absolutes that science discovered.

"All of the best golfers, relative to the worst players, had their hip sway closer to the target at all of those positions [top, impact, and follow-through]," Clearwater adds from GOLFTEC's research. "I think that's something so powerful yet misunderstood. Without your hip sway being closer to the target at the top of the swing, impact, and the follow-through, your odds of hitting the ground in the right spot are very challenging."

Clearwater's research also found the same to be true for the shoulders for top players.

"At the top of the swing, the best golfers, on average, have their shoulders much closer to the target than the worst golfers," he adds. "Hip sway and shoulder sway are both more towards the target with the best golfers."

"If everybody's doing this, then it has got to be important and let's find out why they are doing that," Cheetham adds about absolutes, recognizing the challenge that comes with teaching golfers to retrain their bodies to move in a way that may not match the (inefficient) sequence it had before. "Now you've got a goal that says, we want to sway. We want to lead with a sway towards the target. Then we want to turn. Then we want the rib cage to turn. And we want to work towards that. And it's up to the instructor now to design the training methodology, the incremental training, the drills, et cetera, to get there."

The body's movement is important, and the more (synchronized) movement the better when measuring the best swings. But simply swaying and rotating in perfect harmony isn't a guarantee that your contact will be perfect or that your speed will be optimized. What's the link between the body and those conditions mapped out in chapter 1? The answer: the hands.

In fact, more recent scientific research has led to discovery about how important the hands and grip can be when it comes not only to contact, but to this modern pursuit of speed.

Sasho MacKenzie, like Cheetham, is a leading biomechanics expert and has spent a great deal of time in the sport. A paper he co-authored and published on Golf Science Lab in 2020 tested the four key areas driving force into the swing to create speed for the average golfer:

1. *Increase the linear distance they move the grip during the downswing.*
2. *Increase the force they apply in the direction the grip is moving during the downswing.*
3. *Increase the angle through which they rotate the club during the downswing.*
4. *Increase the torque they apply while rotating the club during the downswing.*

The kinematic sequence is at the root of much of the gains in the fourth area, while the timing of that sequence, MacKenzie points out in the paper, helps with the first area as well. But which force mattered the most? The final conclusion:

Individual golfers do multiple times more linear work, relative to angular work, during the golf swing and linear work also accounts for the vast majority of differences in clubhead speed between golfers. Methods of training that increase the average force applied in the direction of the hand path during the downswing have the greatest probability in generating increases in clubhead speed. From a more practical standpoint, results from this study suggest that for amateur golfers, increasing the length of the hand path is more likely to increase clubhead speed than rotating the shaft through a larger angle.

Simply making the hands travel a farther distance had the greatest correlation to increased swing speed. On the surface, this makes a lot of sense. A short backswing limits the amount of distance—even with great flexibility and a perfect rotational sequence, a golfer can travel the hands and club back to the ball. The simple, linear distance one can move the club improved the opportunity to gain speed. Just another discovery on what can be done to optimize the swing. Find a way to lengthen the arc on which your hands travel.

The hand path helps with speed, but what about the role the hands and grip play with contact itself? That's a frontier that is both as old as time in golf but brand-new in the biomechanical study of the sport. It's been easier to track the full swing—the movement of the club and body —than it is the various pressure points and force impacts that the fundamentals of the golf swing possess.

"Every player seems to have a unique grip force signature as part of their swing just like everybody seems to have a unique swing," says John McPhee, a professor of systems design engineering at the University of Waterloo. John has conducted numerous biomechanical tests of golfers and equipment, advising golf manufacturing companies and publications with the studies he and his students have conducted.

Traditional thinking was simple. A strong grip (bottom hand more open and/or top hand more closed) naturally led the clubface at impact to rotate more closed to the swing path. Anecdotally, strong grippers of the clubs produced more hooking golf shots.

Traditionally speaking, a weak grip (bottom hand more closed and/or top hand more open) naturally led to a clubface that was open to the path at impact. Anecdotally, golfers with weak grips hid more fading golf shots.

Were those traditional assumptions accurate? As discussed in chapter 1, the ball's flight after impact is merely a product of swing path and clubface angle, so should grip matter in those ball flight laws? Both path and angle can be manipulated by the golfer, like Woods and Trevino hitting fifty-yard wedges. The grip doesn't guarantee absolutes, but it does affect the ability of a golfer to manipulate the result and, it was discovered, could influence much more.

"How do we measure the grip? Because how you attach yourself to the club really affects the way the clubhead acts around the impact point and how the clubface closes," Cheetham said.

He conducted a study with Terry Rowles, one of the world's top teaching professionals, where they forced high-level golfers to hit golf shots with different grips and then studied the results. The findings were remarkable.

"When we changed grips, we immediately see the club delivery numbers change," Rowles said. "The consequences on the club were very predictable . . . The path will change. The loft will change because the timing of the release is different. Face angle relative to the path and target will change. And the contact point on the face will change, which is pretty interesting."

It was a major epiphany for Cheetham.

"Effectively what we could see now is there were changes and they were able to adapt, and they were able to change their body swing. What that meant is that it is not necessarily the kinematic sequence changing your swing, it can be something as simple as how you grip the club and how you have the face oriented that now changes your body."

It doesn't diminish the optimized results from a perfectly timed swing, but rather adds another data point for golfers to track. How a golfer puts her hands on the club needs to allow the swing to work better. To an average player, it means more freedom to do what is comfortable, while striving to make it all connect.

Rowles mentioned the release, which is important to note in all of this, how the wrists drive the club to impact in the final choreographed act of this downswing dance. This final act propels the club outward and creates the last bit of centripetal force. Arriving at the golf ball timed to where the angle of clubhead, shaft, and arms are optimized is an important last step in a perfect energy transfer during the downswing.

"You are releasing the angle between the leading arm and the golf club," says Marty Nowicki, a PGA teaching professional who helped develop a training aid called the Impact Snap, which works with golfers to train the wrists to deliver the best release of the club. "At some point near the top of the swing, you're going to have near a right angle, maybe even an acute angle between the leading arm and the golf club . . . What has to happen for that angle to release? The lead wrist has to uncock, or go from radial deviation to ulnar deviation, which is basically cocking and uncocking of the lead wrist, barring what type of grip you have. The weaker the grip, the more you are then going to have to flex the wrist or bow the wrist."

Does that mean everything needs to be in a perfect line?

"That's one of the misconceptions," Nowicki adds. "What you'd be looking for is a lower-case letter 'y' for a right-handed golfer, a lower-case letter 'y' relationship at the moment of impact . . . The club is trailing the leading arm as long as possible. That being said, at about waist high for the longest hitters, it has now caught up to the arm so it's in a straight line. It's in the process of releasing as you hit [the ball]. It is fully released post-impact."

Think of the release of the wrist like a wrecking ball trying to deliver as much force to the wall it is knocking down. If the pendulum it is traveling on (the chain) is behind or perfectly aligned up-and-down with the wrecking ball itself, all of the momentum of the ball has been capitalized and some may have already been lost.

An optimized strike with the most power means that the wrecking ball is still accelerating when it makes impact with the wall. The same is the desire for any golf shot.

The delivery of the club at impact via the release of the wrist should vary from shot to shot depending on the club. The maximum speed desired for a swing with a driver will, naturally, deliver different wrist and release action than a 60-yard, flighted low wedge shot, where a golfer is trying to manipulate the angle of attack and face of the club at impact, but there is still a science to creating solid contact in both examples.

"You really want to keep the lead arm ahead of the shaft until post-impact," Nowicki adds. "That stabilizes the shaft, but if the shaft isn't stable when you hit it, you . . . certainly won't be anywhere near maximum energy going into the ball . . . The geometric appearance of the forearms on the club are always similar."

Seeing this over and over in slow motion has improved the understanding of the body's motion, and studies like what Cheetham and Rowles conducted additionally showed that something as simple as the connection of one's body to the club impacted every step in that choreographed dance down to the ball. While Cheetham is excited to study this more and have the rapid feedback of more data points from a variety of golfers, it's another example of how the uniqueness of one's swing can stay unique, not corrected, so long as it fits the process to optimize the desired result.

What about other areas of fundamentals and setup? Biomechanics has informed the way instructors look at the traditional model of stance and setup. In most cases, it has validated the trial-and-error means with which the game arrived at some setup absolutes. In some cases, it has even brought old habits back to the game.

Like with Lee's front toe swing trigger, more and more elite golfers are moving their feet in various ways throughout the swing to maximize the rotational speed and ground force reactions needed to execute a modern swing. The data and 3-D mapping has validated that work and also driven advancements in injury prevention with so much torque force being driven through the body.

Additionally, much like the grip, the stance has an impact on what the body does. It forces adaptive movement, which can lead to a change in club path and increases or decreases the body and hand's ability to manipulate the clubface in a certain manner. In many ways, it's a data point for the body and what happens next. Synchronizing the body's position at setup with the rotational sequence to follow is important to understand when it comes to mastering the swing.

"I would say that the first half of my career, I focused on where the club was and moving the club correctly. I had some success and that was very helpful," Castro says. "In the second half of my career, I focused more on getting the motion right. Given the choice, I think moving the body correctly is more important than putting the club in the right place."

This leads to a last fundamental on ball position. There is plenty of research available about the right location for a ball to be placed based on the desired shot, but if something can be as connected by both feel and science in the sport of golf, it might be the ball's location in the stance.

The modern driver and ball (lots more to come there) has led to an optimized launch condition that demands an upward attack angle at contact. Forward ball position, in almost all cases, is a necessity there. After that? Echoing Castro's comments, understanding the body's movement and delivery of the optimized swing can inform as much, personally and uniquely, about ball position as anything else. Where to put it has to do with learning where it is struck optimally, or where it needs to be to alter what one needs to do (clubface angle, angle of attack, etc.) for that particular shot.

And maybe that's the full circle-only undeniable truth in all of this—perfection comes from understanding one's own unique way of swinging. There are some absolutes that have been learned, but the body's journey to perfect impact has likely always had a best or better way, tailored by individual ability. Today, with plenty of scientific tools to see it more clearly, each swing can work critically towards its own version of perfection.

"Nothing's really new," Immelman adds. "It's just that with all this paraphernalia we have at our disposal, we can now measure stuff. What can be somewhat disguised on a video camera is now read via data on the clubface [and beyond]. I work with an instrument that tracks your hand path. It gives you over the top numbers, under plane numbers. It gives you swing arm width. It gives you timing from the arms, from start to contact. So, not only can I measure the club with my launch monitor, I can measure your hand path, which is you on the steering wheel. Then I can put something underneath you and it measures the force that you're putting out into the ground, and how you're moving pressure between your legs, and where your center of mass is. Then with my 3-D app on my phone, I can really break it down.

"In terms of using the body more or less, I don't think it's changed very much. I just think we can describe it better."

POWERTRAIN

In the spring of 1997, Paul Stankowski was at the peak of his professional golfing career. He had won back-to-back tournaments the previous year on the Nike Tour and PGA Tour, a dream stretch of golf and a meteoric rise to the game's premier circuit. In February of '97, on his way to becoming a top-25 player in the world, he won for a second time on the PGA Tour and then finished in the top five at the Masters, his best finish at a major.

He was 15 shots behind Tiger Woods.

Speed had arrived in the loudest statement win, arguably, in the history of the game. That pursuit of speed would change the way that golfers looked at themselves as athletes. Woods was still a string bean of a young man but would soon lead a fitness revolution in the sport.

For Stankowski, ask him about his workout regimen in 1997 and he can only laugh. He visited the traveling fitness trailer on Tour one day and hopped on a vertical climber machine to pass the time amongst friends. Most times, he says, it was more about the hang than the work. He recalls spending three minutes on the machine, watched his heart rate soar, and then a player joked about his sudden cardiovascular load.

"I get off and I just walked out the door. I never went back," Stankowski recalls. "But now, everybody's in there, or they have their own trainers, and they're set up in the locker room. They have the physio trailers, the guys are going offsite, they're working out in the hotel gym. They're literally taking their spikes off [after a round], putting their tennis shoes on, and doing some more work, and then getting some body work done and getting a massage. These guys, they're fine-tuned. We didn't have it back in the day."

Thomas Edison's motto was "There's a way to do it better. Find it." That can plainly summarize the cataclysmic shift in golf that took place in about a ten-year window of time from the early 1990s to the early 2000s. Much of that was technological in terms of golf club and golf ball construction, which is coming up in later chapters. The other was one person: Tiger Woods.

Woods was different on almost every level. He won. Brashly. He dominated. Explosively. He conquered. Triumphantly.

To study Woods, even at a young age, within the framework of the first two chapters on contact and biomechanics, he had it all. Repetition from the age of two, coupled with incredible hand-eye coordination made him understand how to manipulate golf shots better than all of his peers. A long, aggressive swing showcased all of the markers of perfect force generation, torque, and kinematic timing. He would have thrived in any era, but he arrived at the perfect time in golf's revolution of understanding and innovation.

Woods would undergo multiple swing changes throughout his career, despite being the best player in the sport, in order to alter club path and gain more control and consistency. He did it all under the meticulous oversight of a mind that understood what his body was still capable of doing. There was a way, he thought, to do it better and he was going to find it.

Tiger didn't *need* to become the strongest athlete in his sport. He *wanted* to become the strongest athlete in the sport. While winning more than a third of his starts in the decade that spanned 1999 to 2009, Woods transformed his body into an athletic specimen. But did he do it in the right way? As his peers (and golfers growing up emulating his game) tried to match his speed alongside rapidly improving equipment, how could they strengthen their own bodies? Like Stankowski on the climber, it was uncharted territory.

"Golf fitness, golf strength and conditioning, as a general topic, it's very much in its infancy," says Jamie Greaves, a former college golfer who now works as a certified strength and conditioning coach and nutritionist. "Fitness has been used in other sports, obviously, for quite a while. It has been in golf a little, with people like Gary Player, but it's really kind of taken off in the last twenty-five years. Because of the Tiger effect, research into it is all quite new."

Fitness has had its supporters through the years in golf, but what kind of fitness was being used and did it have results-improving impact on the golf course? The need for a world-class basketball player to be in peak cardiovascular shape is evident—the sport has high demands on fitness to be able to withstand athletic movements for minutes at a time. Is the same true for golfers?

"Golf is a one-second activity where you are trying to produce as much force as possible," says Mike Carroll, who works with dozens of professional golfers and whose platform, Fit For Golf, is used by thousands of golfers worldwide. "It's one rep at a time. If you think about what most people who delve into working out do, it's usually pretty fast-paced, high-intensity, continuous types of workouts, where they're getting their heart rate up. They're sweating, they're huffing and puffing. But all those things, while they're great for health, general fitness and burning calories, they are not teaching you how to produce more force in a split-second action.

"If you think of people who jump or somebody who's throwing a shot put, or somebody who is hitting a baseball or pitching a baseball, those types of motions where it's a really, really short duration, high-power activity, that is what hitting a golf ball is in terms of physical demand . . . I often think a way of explaining it to people is there's a difference between exercising and having a training program that is designed to enhance the type of improvement that you're looking for."

Like everything else, advanced scientific research and discovery has been at the heart of golf's fitness revolution. Those discoveries created a chain reaction of innovation. What leads to good contact? Repeatable motion and speed. What leads to repeatable motion and speed? Proper kinematic timing, energy transfer, and force.

Now, how does a golfer achieve that optimized speed? For the best trainers and performance doctors in the world, that was the new frontier in golf at the turn of the twenty-first century. It was time to power things up.

The industry took off, with pioneers like Dr. Greg Rose, a chiropractic doctor with a mechanical engineering degree, who helped launch the Titleist Performance Institute in 2003. Today, there are more than nineteen thousand TPI-certified instructors that utilize the knowledge of the body to better inform improvement in the sport. It is a rapidly growing industry with an almost exponential amount of knowledge growth. Like everything else that has been learned about efficiencies and optimization in golf, having the right tools has been critical.

"I think technology with golf club-making, with fitness, with human movement, everything in the industry that helps with performance and recovery, that's the biggest improvement, because everybody has access to it," says Dr. Troy Van Biezen, a chiropractic doctor who was in those early physio trailers on the PGA Tour and still works with professional golfers (and athletes) today on site and in his ChiroSport practice in Dallas. "We're learning things quicker. We're learning things faster. We're getting smarter, and things we used to do back then, we don't do now. That's probably the biggest thing."

What has changed via understanding? If you break golf fitness down into four main categories—mobility/flexibility, stability, strength, and speed—they all work together in unison, and each has its own importance in providing longevity to the player and preventing injury, no matter the skill level. A golfer of any age and body type can improve upon those four areas.

Each is equally important, with flexibility being the largest fundamental held over from previous generations of fitness, while strength and speed have taken on an even larger role in enhancing modern golf performance. But understanding the long-term benefits of strength and speed, while tailoring workouts to the one-second burst needed in golf, is what has changed how golfers exercise.

"As we get a little bit older, basically after our late twenties, early thirties, we start to lose strength and speed and power," Carroll says. "As we get older, it's the fast-twitch muscle fibers—the ones that are responsible for doing the very explosive high-speed movements or the very heavy movements—that start to decrease in size and actually die off quicker. And they can be replaced with slow-twitch fibers if we don't use them.

"If we're only doing cardiovascular-type exercise, and we're not doing anything that requires an extremely explosive, fast, or heavy movement, those muscle fibers don't get trained well, and that's why we lose so much strength and speed. That's why it's so important to have something like jumping, something like throwing medicine balls, something where you're basically moving as aggressively as possible, to keep those muscle fibers working and get our brain used to activating, which keeps those pathways alive essentially. Then, when you're going to swing a golf club, those muscle fibers and those pathways are still there, so the speed is going to be much higher than someone who hasn't trained at all."

Separating what is best for a healthy life and what's best for a perfect golf swing is important in all of this. Of course, getting thirty minutes of cardiovascular work every day is essential for all of the health benefits a doctor would encourage. Mixing that with strength training is important, but there could be even more efficient ways of getting both.

Modern health science studies have found that short duration, explosive workouts (high intensity interval training) help with blood pressure, mental health, muscle retention, and other longevity benefits. You can give your heart a workout, while also strengthening the muscles needed to protect your body and speed up your swing.

Gone from the trailers on the PGA Tour is the climbing machine. In its place, more weights, boxes to jump on, treadmills set for sprints (not marathons), resistance bands, and any activity that can be done to encourage quick, explosive movements. The main goal is to build strength (muscle), but doing so in an explosive way has the added benefit of getting the body accustomed to working in short, athletic move.

"Obviously, you don't jump when you're playing golf," Carroll says. "The reason why jump training is so good is jump training gets us better at pushing force into the ground. If you're someone who is 240 pounds, your vertical jump might not be very good, but you might be pushing a whole lot of force into the ground, and that's what allows them to get so much ground reaction force, the big term that everyone is now using, and that sets you up for powerful rotation.

"Most golfers are obsessed with rotation because when we look at someone playing golf, we see the big turn, which is important, but what sets us up for that powerful turn is how hard we push our feet into the ground vertically. So, if we can improve that push into the ground with our legs, which can be done with exercises like squats and jumping, it makes it way easier to rotate hard and then get our club head speed up."

It's somewhat counterintuitive to think that the ability to drive as much force into the ground is more important to rotation than flexibility, but that's what the science has taught those tasked with optimizing the swing and generating more speed. Speed training is also a muscular development. It's not just about creating the strength to drive the body faster through ground forces, but also the body learning the repetition of fast motion to make it more natural. Building muscles is one thing, firing them as quickly as possible is another.

"It's not good enough to just get bigger and stronger with heavy weights that improve our force production potential," Carroll adds. "We want to learn how to tap into that force at a faster and faster rate. And we do that by practicing moving lighter implements as quickly as possible. Things like medicine balls, bands, and even golf clubs and speed sticks.

"In the golf swing, we don't have enough time to produce all of our strength when we're swinging . . . Rather than applying 50 percent of our strength base, maybe we can learn how to apply 52, or 55, or 57 percent. What this means is that we're getting better at applying more of our force in the timeframe we have in the swing."

Strength and speed obviously aren't the same, but they can both become better when optimized together. You can have strength without speed. You can have speed without strength (and risk injury). Just like the kinematic sequence driving the downswing of the club, the most efficient golfers today have synched their strength with their speed to get the most powerful strike possible to the golf ball when needed. While there is more muscle in golf, it isn't about looking like a linebacker in football.

"Their body is a machine," Stankowski says, reflecting on the rapid change in how young players now look at physical training as a component of their routine. "They're training hard in the off-season. You watch these videos of these guys, during the week or during the off-weeks, getting in there and doing explosive stuff and some speed training and strength training. They're understanding that you have to be stable and you have to have the muscles in your body working properly."

In 2007, the first year the PGA Tour measured clubhead speed across an entire season, the average speed with a driver was 112.37 mph. In 2021, that average was up to 114.42 mph. In that window of time, equipment, largely, hasn't changed in terms of size and weight. Players are almost 2 percent faster because of fitness. In a sport with a number of tiny margins, those are big gains.

"When I first started, golf fitness really didn't exist," Van Biezen recalls. "There was really nothing going on. Guys would do situps, guys would do pushups, or they would do a hamstring stretch, and that was kind of comical at the beginning . . . It's really cool that a lot of these older guys are now in the gym, working out [Stankowski being one of them]. I love that, and it's not just for golf. I think a lot of it is just for quality of life. People just want to feel better as they age. But the missing link I continue to see over and over again, is flexibility."

At the age of fifty, Phil Mickelson became the oldest man to win a major championship in golf when he conquered the field at the 2021 PGA Championship. Three years prior, just after his forty-eighth birthday, he was a viral sensation for a commercial in which he executed a head-high kick dance move that would make most quadragenarians squirm. Related? Absolutely.

Mickelson sought out more speed as he neared his fiftieth birthday and employed many of the tactics already outlined to develop more strength and speed. It was his natural flexibility, however, that allowed both of those areas to thrive and, most importantly, allowed him to still be playing at an advanced age relative to his peers. For Mickelson, flexibility he was born with (and steadfastly maintains to this day) gave him the ability to quickly improve his strength and speed. For others, it may be the other way around.

"If you want to continue to play the game for a long period of time, it's flexibility and mobility," Van Biezen says confidently. "A lot of guys go in the gym and push too much weight and actually lose their flexibility, because they're not actually working on their flexibility. It's a completely different mindset. I love going to the gym and I love getting stronger, but you have to match that. You have to maintain your range of motion with that. Guys get in the gym and pump weight, because they see it on TV and they see it on social media, and then you see them try to swing a club. They can't even do a full back turn anymore. They can't even get to their left side or they can't even load the right hip. That's the things you have got to work on . . . Improve your range of motion in your hips, in your shoulders and in what we call your thoracic, your mid-back. If you can get those three body parts moving better, number one, it takes the stress off your lower back, and number two, it helps you turn better in the golf swing."

This all ties together to the force driving the body's rotation from the ground up. The strongest golfers could load massive amounts of force into the backswing, but the amount of torque energy stored in the body and then transferred through the downswing is directly correlated to how much the body can turn, in all areas and directions. Those limits are determined by flexibility. While everybody can make a swing, how far they can stretch that swing is a limit worth pushing for optimization.

"The more flexible, the more I can turn, the more elastic energy

Dutch professional golfer Daan Huizing warms up before a tournament on the DP World Tour in February 2022. Increased focus on flexibility movements and rotational stretching is a part of almost every golfer's pre-round routine. *Getty Images*

I can create, the more speed I can generate to my club head," Van Biezen says.

Van Biezen and other chiropractic doctors were also at the forefront of learning about the biomechanics of the golf swing through three-dimensional mapping and imaging. Van Biezen was heavily involved in the work of Gears, another leading biomechanical sports company, which mapped out areas that led to strain. In the world of the doctors, achieving personal bests and avoiding injury go hand-in-hand. Learning what doesn't work for the body is still taking place.

With all of the added muscle and speed, there is a natural toll on the body. Past accepted practice thought that a stance with the backside more protruding was a good posture. Many other desired swings involved a stacking of the body in a way where hips weren't parallel to the ground. All of those setups, when examined with three-dimensional data and injury tendencies, showed an undue amount of stress on the body.

"We want the spine to be neutral, and the more neutral, the stronger position it's in. And that's just a good base to start from," Van Biezen says. "As we get older, guys in the fifty-five [year range], into the sixty-five, seventy-year-old range, flexibility is obviously keeping them healthy and keeping their back feeling good."

It's impossible to discuss injury and longevity without returning to Tiger Woods, whose play (and physique) led to much of the understanding and exploration of the golfer as an athlete. He was also was the most public example of what can go wrong. Woods swung with such ferociousness that his use of ground forces is its own case study. But hundreds of thousands of violent swings drove that energy up through his back and legs. His survival as a golfer was dependent on the stability muscles (glutes, obliques, abdominals) in his body allowing him to make such powerful, rotational moves. Once his fitness routine wasn't fully invested in protection, in addition to optimization, things started to break down.

Wright Thompson wrote one of the most detailed breakdowns of Woods's tumultuous personal and professional journey in "The Secret History of Tiger Woods," still available on ESPN.com. In it, he logged the many physical feats Woods took on to better connect with his father, Earl, who passed away in 2006. The elder Woods, a United States Army Special Forces veteran, produced one of the most impactful father-son relationships in the history of sports, and Tiger channeled much of his grief over his father's passing into experiencing the workouts of a special-ops military life.

Four-mile runs in combat boots, martial arts sparring, and combat simulations became a part of his routine in 2006 and 2007. He was getting older while doing more work that, by today's understanding of golf fitness, didn't seem conducive to the fast-twitch muscle maintenance needed. Whatever personal and psychological benefits the workouts may have had for him, it became impossible to deny the physical toll they took on him for the rest of his career. He beat up his body and his body eventually beat back.

There were lessons taken from that. Tiger was the standard, good and bad, in how golfers learned to deal with their bodies and protect it for the marathon of a career. Unlike football players, who build brute strength and muscle mass to survive a finite period of time being bludgeoned, golfers train for strength, speed, and flexibility with an eye on still being able to move effectively decades later. That takes a commitment to a core holding it all together.

"This is where the fitness world in golf—if you look at fitness in golf compared to other sports—it's still new," Van Biezen says. "We're still learning what works, what doesn't work. We're collecting more and more data. But what a lot of people have access to, looking from the outside in, is they see how a lot of these guys train. They do Olympic lifting, they do powerlifting. They see Brooks Koepka, Bryson DeChambeau, guys like that.

"But what they don't see, which is not the sexy part of fitness in golf, is building a mobile stable foundation. It's boring, but learning how to control your body in different phases of movement, in different planes of motion. Exercises like side planking, front planking, bird dog, dead bug, just basically getting the body building up that stable, mobile foundation."

Just like the kinematic sequence relied on the timing of four areas of motion to get the downswing in perfect rhythm, golf fitness relies on the perfect coordination of those four areas to get better or simply sustain. And just like the downswing in motion, one area out of order can break the whole system.

"I didn't have all the data that the guys have now," says Arron Oberholser, PGA Tour winner, who was inspired by Woods, a competitive peer, to get stronger in his game at the same time. "I didn't have all the information that they have available to them now. I just basically got in the gym and got big (bench-pressing 300 pounds) and thought I was doing the right things to get speed. I gained zero distance."

How does all of this added work the body is doing get fueled? With golf still very much at the infancy stage of athletic study, the area of golf nutrition does not have a number of large-scale studies from which players and coaches can develop plans, but connections with other sports and some early research are leading to larger discoveries when it comes to fueling golfers as athletes.

"Most people would assume there's not much activity going on other than maybe the swing itself," Greaves says. "The research that is out there on this has shown quite a wide variation, but your typical golfer might be burning upwards of a thousand, maybe fifteen hundred calories during a round of golf . . . It's certainly more than what most people think. Because there's this underappreciation of what we need to eat and drink on the golf course, I would wager that 90 percent of golfers are not eating or drinking enough on the golf course."

Before golfers even get to the course, prepping for the added demands of an increasingly athletic sport has shifted the idea of nutrition to a different level. With the strength training discussed earlier, high-protein diets become more important for golfers to maintain strength and build muscle.

At the highest levels of professional golf, with days consisting of eight to ten miles of walking, keeping muscle weight on is a challenge. Greaves figures golfers are consuming protein at almost a 1:1 ratio of grams to pounds in conjunction with most modern strength programs. But what about the diet makeup to get ready for a round?

General rules on nutrition aren't common. Think about all of the various diet plans that exist in the world. There is mixed science as to what is the perfect formula of food and nutrients to put in the body. Much of it is results-oriented. The rest is personal preference and body chemistry. If there is a universal thought as it relates to proper nutrition during a golf round, it is to not overwhelm the body at any moment. For the body to react at its optimum speed and rhythm, it can't be sluggish.

"Most people and most golfers, they don't really want to eat a meal out on the golf course, so we're going to be looking for getting three to four snacking opportunities," Greaves adds. "A really nice way to break it up would be a snack midway on the front nine, maybe holes four to five, a snack at the turn and then a snack kind of midway through the back nine, so [holes] 13 or 14. Three good opportunities, getting in a few hundred calories each time. That's not going to feel too overwhelming for the golfer . . . because the last thing you really want to do is take in a huge amount of food and then start to feel really sluggish, because obviously that's not going to be good from a performance standpoint either."

What you eat. When you eat. How you stretch. Building your strength. Training for speed. Establishing your core. All of them, together, allow the body to do its best work. Ignoring those? As Van Biezen closes:

"If your body's working against you and things you want to do in your golf swing, you don't have a chance."

4

SIX INCHES

Golf is a game of inches. The most important are the six inches between your ears.

—Arnold Palmer

On a Saturday in July of 2021, Collin Morikawa was still just twenty-four years old and had been a professional golfer for barely two years. He had already won four times as a pro, including one major at the 2020 PGA Championship, and found himself on the brink of history again, one shot behind the leader in the Open Championship, golf's final major of the season.

It hadn't been a perfect day, with a slow start costing him a shot at being the outright leader of the tournament. Royal St. George's Golf Club was playing difficult, a test for even the most veteran of golfers. Prior to that week, Morikawa had played links golf—a style attributed to the demands of the land, conditions and architecture of older courses throughout the British Isles—just once, and it was a run-of-the-mill performance just the previous week.

Yet, here he was again near the top of a leaderboard littered with the best players in the world. How had his star shone so bright so early? How would he contend with realities facing him going into the final round? How would he compartmentalize his mistakes under the pressure of needing to be nearly perfect the next day to win?

"I wasn't hitting that poor of golf shots, it just wasn't turning out great," Morikawa said in his post-round interview on that Saturday. "No matter what happens tomorrow, I know I produced good golf shots already this week and I'm capable of it. I just have to stick to that and believe in the process."

The process. A concept that Morikawa had drilled into his mind since before he was a teenager. At the age of eight, Morikawa started working with PGA professional Rick Sessinghaus, who also happened to have a doctorate in sports psychology. Their work through the years certainly honed Morikawa's world-class ballstriking (in 2021, he was the best iron player on the PGA Tour), but it was their work on the mental side of the game that prepared Morikawa to be ready to handle any situation at the very beginning of his professional career.

Morikawa didn't make a single, critical mistake in that final round. He shot a bogey-free, 4-under-par 66, led by at least two shots throughout the entire back nine, and won his second major.

"You can't worry about the score," he said afterwards. "I had to worry about every shot. Can I execute every shot to the best of my ability? Some we did, some we didn't, and then you move on. We can't control what's happened. I really looked at that as just focus on every shot, how do I see what the best shot is possible and try and do my best from there."

His best made him the first golfer to win two different major championships in his first appearance and became the first golfer in nearly one hundred years to win two major championships in eight or fewer starts. It took physical skill to make history, but getting his brain to maximize that skill is something that has plagued golfers of all levels since the dawn of the sport, and is widely studied and discussed today.

With a golf swing that takes only one second to achieve, we have established the elaborate science in understanding the incredible symphony of moves and adjustments working together to create the perfect swing. The conductor of that symphony is the brain. The messages firing from the brain tell the body what to do. A sharp mind is necessary to guarantee peak physical performance.

Conversely, a weak mind can destroy a golf swing or golf shot. A pressure putt that is pulled sharply to the left is viewed as a mental breakdown, when the reality is that it was a physical response that led to the miss, as directed by the brain. The mind and body must work together in harmony for the golfer to succeed. In some cases, this requires intense focus. In others, it may require a calm clarity or a clear mind. In modern sports, this has led to more and more work from sports and performance psychologists.

Dr. Debbie Crews has spent more than thirty years working towards a greater understanding of the brain's involvement in the golf swing. She studies how behavioral, cognitive, and psychophysiological variables lead to optimal performance. Specifically, she has studied attention in golf, examining brain and heart activity, both of which are indicators of attentional focus.

For Crews's research, it again comes down to one second. This time, it's not the one second it takes to swing the club, but the one second prior to the brain sending the message to start and complete that swing. Crews has discovered through her research that the pathway to better success was through a balanced brain.

"In golf specifically, I kept seeing this pattern of synchrony," she told the Golf Science Lab podcast in 2017. "As the left side of the brain is quieting in the last second right before people move, the right may become slightly more active, but what you achieve is balance or synchrony in the brain, and it was the last second of data that was predictive of performance."

The left side of the brain is tasked with many of the basic functions of the body. It is the logical side, the mathematical calculator of what needs to happen. Typically, it functions in a very linear way. For a golfer, this may mean understanding all of the variables that need to be combined in a given shot. For example, it sees a putt of 15 feet, calculates the force needed to hit that putt and the direction the ball needs to start rolling in order to use the contours of the green to adequately reach the hole in conjunction with that applied force.

The right side of the brain is the artistry of that calculated motion. It is tasked with visualization and imagination. It also is the source of rhythm in the body. Typically, it functions in a more abstract way. For a golfer, this may mean coming up with more creative solutions and feeling what the body needs to do to exercise the shot. In our example, the right side of the brain will deliver the pace in the arms with which we deliver speed to the putter. It will look down the line of that 15-footer and visualize the path the ball will take to get to the hole based on that slope calculation the left side factored.

Much like a swing that is synchronized, Crews's research shows that the brain must be in total balance as well to drive the body in an optimized way. The pre-shot routine and every other non-swing moment of the round of golf can have all sorts of variety between left and right brain activity, but when it comes to that second of time before the club starts the backswing, left and right need to be in harmony. Typically, for most golfers, that involves transitioning from left brain thoughts to the right brain to balance it all out.

"In the research, when we asked those people who are performing well, 'what is the last thing they are thinking about before they start moving,' the typical response is either 'target' or 'feel,'" Crews also said on the same podcast with Golf Science Lab. "If they are doing semblance of the target, that's going to be a right-sided activity (visualization), and if they are in feel mode, that's going to be a right-sided activity (rhythm). They can use cues to help them. Anything,—timing, rhythm, balance feel—if they are taking in information instead of trying to micromanage the emotion, that is going to help them in terms of getting that synchronized place in the brain."

This is the sweet spot now for the most sophisticated of golfers. With so much information available to achieving success, swinging a golf club is more left-brained than ever. It is a scientific calculation of force, sequence, and body control, and that's after calculating club selection, shot type, weather variables, and other countless data points in the mind. It can be an overwhelming physics problem. What must happen, the research shows, is for that left side to shut off enough to engage the right side of the brain in synchronization to take the club back and make the swing. It's now a two-second dance of mind and then body.

The technology and data that Crews developed became OptiBrain, an app that involves the user wearing a brain monitor that works in conjunction with one's smart device. With it, and other similar technologies, coaches can map the baseline brain activity of a golfer (some people are more naturally dominant on either side of the brain) and then compare that to what the brain waves are doing leading up to the swing. The apps track not only when the brain is balanced in activity, but also log which brain patterns led to the best desired results with the swing.

Some golfers might call it the "zone," that moment of absolute clarity. It's easy to describe the feeling after the fact, a product of mental calmness and flawless execution. With a brain monitor, the "zone" was now alive, an image of a balanced brain.

In a study of the effects of her product—which played music at varying volumes based on brain activity to teach golfers how to find that moment of clarity (balance) before initiating the swing— Crews and her team saw a 16 percent increase in putting success (putts made) versus a control group of physical practice only. There was real science to indicate that arriving at that focused moment of brain balance will lead to a better swing.

Were there tricks to achieving that balance? Crews links the brain and the body together in both directions. Working with athletes on balance boards, she could see how the brain was balanced when the body was solid as well. The idea of mapping brain activity and biomechanical measurements together is still very new, with little substantial research, but there is a natural correlation, it would seem, between the strengthening of stability muscles and mental balance.

Another key to success in achieving a balanced brain brings us back to Morikawa and that magic word: *process*. Crews's research of high-level golfers found that the best ones—those who could control their brain and find its synchronization—were able to exist in the present, not dwell in the past, and move forward to the next task seamlessly. Again, to Golf Science Lab:

"In terms of brain patterns, [high-achieving players] don't question, 'will this go in' or 'will I perform well today' or whatever it happens to be," she said. "For some reason, that seems to be missing in those people. They just perform and they are always moving forward."

When Mickelson made history at the 2021 PGA Championship—winning at the age of fifty—he certainly had put the physical work in to compete at the highest level. But it was his mental commitment for four rounds that week that he initially credited with victory.

"I've not let myself think about the results until now, now that it's over," Mickelson said immediately after the win. "I've tried to stay more in the present and at the shot at hand and not jump ahead and race. I've tried to shut my mind to a lot of stuff going around. I wasn't watching TV. I wasn't getting on my phone. I was just trying to quiet things down because I'll get my thoughts racing, and I really just tried to stay calm. I believed for a long time that I could play at this level again. I didn't see why I couldn't, but I wasn't executing the way I believed I could, and with the help of a lot of people . . . I've been able to make progress and have this week."

There are two major elements there that deserve exploration. The idea of staying in the present is nothing new to psychology, especially among analysts in the sports science world. But in golf—a sport of repeatable, unguarded, individually controlled moves—every stroke is its own, self-contained battle. That routine should be exactly the same. To confuse it with additional information, like potential outcome or previous result, is only adding noise to a brain that needs to transition away from extensive calculations and get more to the right side.

"It's about leveling out that emotion after a made putt or one hole and then going back to the next tee shot and realizing that that was the last hole," Morikawa adds. "You've got to worry about this next hole, kind of this mini-match."

The more we understand about the brain, the more the science validates the need for routine. Process has become another word for routine for golfers at the highest levels. The more repeatable everything about the shot, hole, round, and week becomes, the more rhythmic all activities are. That rhythm is what the right side of the brain demands. Cloud that process with unseen variables, and you are asking the left side of the brain to come in more often to calculate what is taking place. It then requires more work to return to that process, leading to both a lack of balance and mental fatigue.

"After you hit your shot, you have to let that go and then enjoy the walk to the ball," says Gregg Steinberg, a sports psychologist who has worked with golfers at all levels. "You have to think of concentration like a mental energy. If you have the gates wide open, then you're going to be drained at the end of the round. If you only open the gate a little bit right before the shot, and then open it wide open at the shot, and then close it again after the shot, then you're going to have a lot more mental energy for the whole round."

Mental energy is the same as physical energy. It requires fuel and stamina from the body. In fact, studies prove that the more active your brain is in cognitive thought, the more glucose it requires and burns. The more you can shut out overthinking, literally the more energy you conserve.

Conditioning the brain has the same value as conditioning your glutes. Both are going to be needed to be strong from beginning to end of a round. If they begin the round already tired, how can one expect to have balance, mentally or physically?

All of this focus on process has led the best golfers to adopt the same routines. Gather information, select the shot, select the club, and then start the switch from left brain to right brain. Visualize the shot, step into the stance, same number of practice swings/waggles, and now the right brain has balanced the left. Swing.

This idea of process has trickled down to more than just the golf shot.

"The last thing you want to probably be doing on a tournament week is start in-taking food that you are not familiar with, trying to change too much," Jamie Greaves says. "You always want to be sticking, on tournament weeks, to things that you're familiar with. I talk about that from a gym setting, as well in terms of on tournament weeks, sticking to exercises that you know, exercises you're familiar with. Don't start putting in new stuff in a tournament week because the body, whenever we put a new stimulus onto it, it can sometimes react slightly differently."

What about the yips? Countless golf games have succumbed to the death grip of the yips. Some players can't take the club back. Others can't start the downswing. Many can't finish a putting stroke, leading to short misses. It's all mental, and it's not, surprisingly, all that complicated.

Dr. Crews has conducted a number of studies on this, some of them in conjunction with the Mayo Clinic looking to link the yips with physical issues. While some versions of the yips have neurological ties to failure (think sudden cramps), a vast majority of the observed yips come back to mental focus. Additionally, while anxiety can amplify the severity of yips, Crews found through studies using beta blockers that anxiety is not the direct cause of yips.

What she did find, whether they were consistent or intermittent yips, is that the left side of the brain was vibrant with activity when a player was experiencing a moment of the yips. Simply put, the yips were caused by focusing too much on the future results.

Past failures or struggles are magnified in the mind of the player, leading to negative thoughts that the next stroke or swing could lead to a similar result. A golfer was not present; instead he was stuck analyzing the past or fearing a future he couldn't control. He wasn't trusting the process, investing fully in calming the mind and giving it over to the right side.

Crews anecdotally pointed out that most golfers don't experience the yips while practicing. Why? Because the result of a practice shot or putt doesn't carry the same importance as one on the course. The brain is able to balance itself easier in practice than in on-course application.

Mickelson's second key to mental success at the 2021 PGA Championship was belief. One of the greatest golfers of all time—a man with nearly 60 professional wins and a half dozen major championships—needed to believe he could win. It's a doubt that golfers of all skill levels struggle with.

The brain science behind belief or confidence is applied in all walks of life. When we experience something positive, our prefrontal cortex lights up and our whole body feels good. Confidence quite literally gives us a rush of positive energy. That energy can be channeled into whatever we need, like focus and motion. It's not noise, but rather clarity. It can also be remembered and conditioned.

The neurons in our brain are conditioned to remember everything. The brain learns as it experiences various stimuli. The more the brain can experience positive moments, the easier it becomes to recall that euphoric energy when needed. It has stored positive energy. So, when a golfer needs to avoid negative memories or the fear of potential negative results, having more positive memories (confidence) stored up in reserve is the key to controlling the mind, getting into the process, and balancing the brain.

Many of these positive memories should be developed on the golf course. For example, a golfer with the yips should focus the mind on all of the three-footers that have been made in the past, not missed. Those memories flood the brain with positive energy, not doubt or worry. Remembering specific mistakes and using those memories to project future failure keeps the left brain in charge. Recalling the success of countless made putts leads to the visualization of the next made putt, which helps transition to right-brain thinking. The same mind trick can be done while practicing as well.

"I think practice drives so much," says Ben Crane. "If you're practicing the wrong thing, or you're practicing something that's challenging for you, and you're not very good at it, you get frustrated and then you have a heightened sense of, 'I'm not very good at this.' Now I'm in a worse [mental] state going out to play golf."

Crane worked with his mental coach, Lanny Basham, to develop a way to improve his game while also not stressing his mind and inviting doubt.

"Sandwich training just means to practice something you're good at," he says. "I'll start off with my putting." (Crane was the top putter on the PGA Tour multiple times.) "Then, say I'm struggling with mid-range wedges, I'll practice my mid-range wedges, and then maybe I'll go back and I'll finish my practice with putting 3 to 10 feet, or something like that. Now, I leave the course feeling great. You're putting something that's challenging for you in the middle of two things that you're very good at. That is a great way to train."

Away from the course, the power of positive thought is another universal practice that has become more of a tactic used by golfers of all levels to condition the mind. This isn't just tied into living a positive life and having positive experiences. Successful golfers will journal activities, tracking positive progress or moments, while reflecting on what worked and what didn't. This concentration on mindfulness is another tactic in training the brain to remember the positive (not just good moments, but positive outlook in overcoming negative moments) and ensure a more likely positive outcome on the course.

Bringing negative energy to the golf course, or keeping it on the golf course, only forces the brain to focus on too many distractions. Negative ideas naturally pull a golfer away from the process, flooding the left side of the brain with too many things to think about.

"The attitude of gratitude is that when you're under pressure and you're still grateful that you can play golf, it alleviates the pressure," Steinberg says. "A lot of times people are in the moment and they're having a bad round and they don't have that attitude of gratitude. It creates more pressure."

"I've always been intrinsically motivated because I love to compete, I love playing the game," Mickelson also said after his PGA win. "I love having opportunities to play against the best at the highest level. That's what drives me, and I think that the belief that I could still do it inspired me to work harder."

This final point from Mickelson reflects another important element to building confidence, positive thought, and an optimized mind for golf. It is a major component in the mental platform that Rick Sessinghaus has developed through the years and is optimized by Collin Morikawa today. Motivation was the first principal discussed in Sessinghaus's book *Golf: The Ultimate Mind Game*. An excerpt:

> *When we turn our attention to what we want, we become more focused, motivated and use our time more wisely. Golfers who make the biggest improvements know specifically where they want to go and in what time they want to get there. Components of motivation involve having a dream, a vision on how to accomplish the dream, the desire to keep working*

*despite the obstacles, and loving what you do. **Motivation is the drive that leads to action being taken toward a desired goal.***

Knowing how the brain operates via the studies that Dr. Crews conducted, this concept of motivation makes perfect sense in terms of activating a balanced brain. Motivation to achieve something never done before (think proper goal setting) triggers visualization and positive imagination, right-brained functions. Critically thinking about the process necessary to achieve those goals gets the left brain and right brain working together to achieve something new.

Morikawa has talked a lot about this, about setting lofty goals early and working towards them. Even golfers at the top of the sport will look for small ways to get better. Perhaps it's a tenth of a point better in a statistical area. Perhaps it's adding five yards of length off the tee. Striving for a new height drives positive energy in the mind.

"The idea is what we call a mastery mindset and a mastery mindset is 'I'm always working on self-improvement,'" Steinberg says. "If I'm always working on self-improvement, it's fun. The best way to explain it is when you're playing a sport and you plateau, and you're not getting better, you lose motivation. But if you always can get better, you keep your motivation to practice and compete."

Believe, focus on the positive, create a routine, and stay motivated. Those tactics are part of a mental strategy more and more golfers employ to help balance the brain and quiet the mind in pursuit of that perfect zone before the shot.

Another important understanding that is becoming more measured is the fact that the brain, like any other part in the body, needs to be taken care of to work at peak function. In the same way a high-level golfer wouldn't hit the golf course without food in the tank to provide energy, or protein in the diet after a strength workout, the brain's physical needs must be met. The first, would be fuel.

"You only have to be 1 percent dehydrated for it to start affecting cognitive performance," Greaves says. "When we are in stressful situations, what tends to happen is we tend to just forget basic things. You'll start seeing people walk faster. Maybe they have this habit of between every green and the next tee they take some water on board, but when they get into that stressful situation, they're so pumped up they forget to do that. They tend to forget basic things. We say to ourselves, 'Why did we do that? What are you doing?' We never link it back to, 'Maybe I'm just kind of fatigued. I'm under-fueled. I'm under-hydrated.' We always think it's a mental error, but what did that mental error stem from? Maybe it was because we hadn't fueled ourselves properly."

Greaves mentions fatigue, which can be physical from a lack of proper body conditioning, but to the brain, much of that fatigue comes from simply being able to give the mind a rest. Proper sleep and rest habits are critical to a solid mental approach.

Monitoring those patterns and generating data individually and globally has become big business in the sports performance world. Companies like Whoop give players daily feedback on heart rate to see how much stress their body is under and how intense certain pressure moments can be. In the framework of the mental game, it works twofold to provide immediate data feedback about what the body and brain both need depending on the workload of each day and/or moment.

"It will give you, 'Here's what you did today and here's what you need tonight,'" says Mike Thomas, PGA teaching professional and father/coach to his son, Justin, who is a PGA Tour Player of the Year, major champion, and early adopter of Whoop technology. "What I try to tell my students about the reason why I use a heart monitor is you are feeling the end result. Your heart rate was rising prior to you feeling it . . . I know the anxiety during [making a double bogey], but the problem was your heart rate was going up on the two holes prior to that because you missed a short putt, you made a bogey, and your heart rate went up. You are asking yourself to perform at an ideal level when your heart isn't at that level and can't cooperate at that level."

Leveraging these new technologies has allowed more players to understand how their bodies respond to stress and work on solutions to manage those moments that elevate heart rate, which could lead to mistakes in the rhythm needed to execute a shot. In fact, it was Thomas's own data on Whoop that showed his ability to lower his heart rate through a consistent pre-shot routine (the process) to execute shots under the biggest moments of pressure. Dr. Crews's own research with heart monitors has also found that the most successful golfers see a heart rate drop of approximately six beats per minute in the three seconds right before the golf swing, validating the science behind quantifying the "zone," that moment of focus and balance.

At night, these new technologies also measure sleep efficiency, encouraging not just a controlled quantity of sleep (there is both too little and too much sleep), but also consistent nightly rhythms of when to wake and when to go to bed.

Perfect sleep can help drive mood, which helps lead to an easier grasp of the positive outlook work golfers are trying to achieve. Sleep also has numerous physiological benefits, from workout recovery to strength-building and immune support, but its impact on brain function may be its most important byproduct.

A 2004 study at the University of Pennsylvania found that adult subjects who were deprived of two hours of sleep per night (down from the medically recognized standard of eight hours to six hours) over the period of two weeks encountered the same level of cognitive performance deficits as subjects who were sleep deprived (did not sleep) for two days. The residual buildup of bad sleeping habits could have the same damaging effect on brain function as somebody pulling an all-nighter.

When the brain is required to perfectly time a rapid sequence of movement in one second to hit a golf shot, it can easily be inferred that a lack of positive sleep will dramatically impact the brain's capacity to conduct that sequence perfectly, especially over the course of a round with seventy or more motions.

According to research by Dr. Marcus Raichle, the brain represents just 2 percent of the body's weight, but accounts for a whopping 20 percent of the body's energy use. Is it fair to say then, in golf, that the mind is 900 percent more important than the rest of the body? Perhaps so.

DATA AND DECISIONS

The lowest score wins but you want to know how a player shoots the score that he does.

—Mark Broadie

At the end of 2014, Rory McIlroy was the undisputed best player in the game of golf. He won four elite tournaments that season, including two majors, ending the year ranked number one in the world by a wide margin. His PGA Tour scoring average that year was 68.8, leading the tour.

The very next season, Jordan Spieth overtook McIlroy as the top player in the game, winning five times, including two majors, ending the year ranked number one in the world with a higher ranking average than McIlroy the previous season. His PGA Tour-scoring average that year was 68.9.

By traditional standards, McIlroy's 2014 and Spieth's 2015 were practically identical. They achieved the same results in terms of winning, especially in golf's biggest events. Each became the best player in the world, and over the course of dozens of rounds of golf, had nearly the same scoring average. Were their years really that similar?

McIlroy was an average chipper of the golf ball in 2014 and did not rank as one of the top 40 best putters on tour that season. Spieth was top 10 in both categories in 2015, but was not a top-10 player in 2015 in either driving (Rory was first in 2014) or approach play.

Same scores. Same results. Yet different ways to get there. And known with far more detail thanks to the modern advancement of golf statistics and analytics.

In the end, the score is all that matters. You are defined by the total number of strokes it took to get the ball in the hole 18 times in a round. (Unless it is match play, which seems like a different book for a different time.) Improvement, however, couldn't come directly from looking at a score. With so many aspects of the game, learning about how a golfer arrives at a particular score is a statistical science that has redefined how the sport is measured.

Throughout most of the twentieth century, golf statistics were a basic collection of absolute totals. A round of golf (so many strokes) was made up of the total number of fairways hit in regulation (first attempt off the tee), greens in regulation (approach shot hitting the green on the first shot on a par three, by the second shot on a par four, or by the third shot on a par five), total putts, scrambling (if a green was missed, does it take two or less shots to hole it out from there), and a few other result-oriented data points.

These binary measurements provided no real context to how a round of golf unfolded. Is a ball in the fairway 200 yards from the hole the same as a ball in the same fairway 125 yards from the hole? Traditional stats rewarded both in the same manner; yes! Curious supporters of the game wanted more when tracking their rounds.

An early creator of advanced golf statistics was Peter Sanders. Intrigued by the capabilities of computer modeling in the 1980s, and an avid golfer, Sanders wanted to better quantify how golfers of all skill levels reached the scores they shot. He partnered with golf schools at Pinehurst and with *Golf Digest* to track the rounds of more than seven thousand golfers and came to prove what he already intuitively knew.

"All of the traditional statistics talked about positive stuff, but It wasn't the positive stuff that really controlled scoring. It's the relative frequency and severity of errors," Sanders found. He developed a system called Shot by Shot, a statistical tracking software for all golfers, and he continues to advise some of the world's best players in monitoring their stats. Beyond calculating the value of those errors, there was still work to be done to figure out how each shot impacted the final score.

The goal of advanced statistics is to break down the round into what went right and what went wrong, using a baseline of a certain skill as the measuring point from which all strokes could be compared. Once you have enough data to see how a player generally achieves a certain score, you have an advanced statistical model to compare others to.

The most celebrated and successful advanced statistical model was developed in the early 2000s by Mark Broadie, a business professor at Columbia University. His system would come to be known as strokes gained.

"What strokes gained does is it translates everything into units that make sense, which are strokes, because the lowest score wins and you want to know how a player shoots the score that he does," Broadie says. "The problem with traditional stats is, say, driving distance, you know that longer is better, but it's not really clear if you hit it a 10-yard-longer drive versus missing a 20-foot putt, how do those things compare to each other? Whereas, if you measure everything in units of strokes, then it becomes really clear how a player is gaining or losing relative to the field."

So how does it work? Let's say on a particular course that a scratch (zero handicap) player's average score is even-par 72. That doesn't mean that every hole is an exact, round number average for every round. The most difficult hole may be a Par 4, 405 yards in length, that plays to a scoring average of 4.3, and the easiest hole may be a Par 5, 475 yards in length, that plays to a scoring average of 4.7. Add up all of those averages and you get the average total score.

For the example of a scratch golfer playing that tough hole, she scores 5 in a particular round being measured. That would mean she is 0.7 shots above average on that hole for that round. She has lost 0.7 strokes compared to the average of a scratch golfer. Beyond the total score performance on that hole, what makes up those five strokes on the hole and how does that compare to other averages of a scratch player? Time to look at the hole from a strokes gained basis for each shot.

Say this hypothetical golfer hit a great drive, right down the middle of the fairway with above-average length. Using a baseline of previous tee shots, the data would indicate that a drive of 225 in the fairway is the median point where a scratch-level player would typically finish the hole with an average score of 4.3, our baseline total hole average. In this example, she hits her tee shot 235 yards. This added distance shows that the average score on the hole for a scratch player from this location is 4.1. She has gained (theoretically at this point in relation to the actual, total score) 0.2 shots before hitting her second shot. Because of the strength of her off-the-tee shot, it is expected for her score to be better than the 4.3 average. One shot into the hole, she is **+0.2 strokes gained off the tee and total**.

Next comes the approach shot, which will be measured from a baseline of similar approach shots on this hole and all holes. This scratch player now has 170 yards into the green. The baseline, as noted, from this distance is 4.1 on the total score, or—since she has already hit one shot from the tee—3.1 shots on average to finish the hole from this point forward. Working backwards from data on putting we have from the green, staying "on average" would be a shot hit on the green approximately 40 feet from the hole. The golfer in our example, however, misses the green, in the rough, 80 feet away. Based on the data, this shot loses her 0.6 shots. Two shots into the hole, she has +0.2 strokes gained off the tee, -0.6 strokes gained approach, and a **-0.4 strokes gained total**.

The chip is next, and it is a great one, hitting it to just five feet. Compared to expected results and data averages, this above-average effort is worth 0.4 strokes around the green. Three shots into the hole, she is now **+0.0 strokes gained total**.

Finally, it's the putt, and it's only five feet away. A scratch golfer would be expected to make this putt roughly 70 percent of the time, with an average of 1.3 putts necessary to get the ball in the hole from five feet. She misses, however, requiring two putts on the green to complete the hole. She writes a 5 down on the scorecard, having lost 0.7 strokes on the green, the most costly area of play in terms of score impact. On this one hole, her **strokes gained total is -0.7.**

Mathematically, this adds up. The hole had a scoring average of 4.3, and her score of 5 was 0.7 higher than average. She has lost strokes in comparison to that scratch average by making a bogey. What the advanced statistics have done is offer a more concentrated glimpse into how she arrived at that bogey. The driving and chipping were good and above average, but the approach play and putting were what cost her a chance at a lower score. This one hole is a small (hypothetical) data set in understanding the overall game of this golfer, but over the course of 18 holes, or a month, or a year, golfers can now see a breakdown of different areas of the game and how it positively or negatively impacts their ability to score. Millions of data points also mean that there is a statistical model for all handicap levels, allowing for rapid improvement and comparison. There is certainly more than one route to that simple number on the card.

Broadie's book, *Every Shot Counts*, lays out the creation of the system and the baselines for all golfers, and is considered one of the most influential books in the sport's history when it comes to the statistical revolution. Broadie's data was broad, and the most thorough of it came from the best.

The PGA Tour officially started tracking statistics in the 1980s, with a rudimentary system of those standard statistics like score, fairways and greens hit, and putts made. In the beginning, it was done with paper and pencil. By the end of the twentieth century, mobile data devices allowed for more real-time reporting and live scoring, but it wasn't until the early 2000s that the system became high-tech.

Traditional stats offered no nuance in the wide-ranging venue of a golf course. Two golfers could hit a fairway but be 50 yards apart. Two greens in regulation could have 10- and 40-foot putts, but be measured as the same accomplishment. The PGA Tour had to introduce a distance variable to their measurements.

This began with the introduction of volunteer-manned lasers in the 2004 season. Every shot hit by a player in competition was measured with a laser, which relayed distance data on every shot back to a centralized database. Suddenly, players (and all constituents) could learn about exactly where every golf ball wound up.

Today, while lasers are still used for measuring the distances of golf balls in landing areas like fairways, the tour employs launch monitors on tee boxes and three triangulating cameras above every green to measure the golf ball in three dimensions, securing that distance data variable and countless other data points for how the golf ball is acting and reacting for each player in tournament play.

"The biggest difference in what the viewer or spectator sees now is where the ball is located, which we never had," says Don Wallace, the senior director of ShotLink, the PGA Tour's data gathering program. He has overseen the logistics of the program from its infancy. "Now we have that real-time nature of having a location, which gives us distances . . . and it's consumed in many different ways."

Enter Mark Broadie. That first data set in 2004—and every year since—allowed Broadie to work with the tour and form their own baseline on what constituted a great, average, or below-average golfer in various areas beyond score and those outdated stats. At first, it was used simply to develop the strokes gained model for putting.

"That was the greatest need," Broadie recalls. "The PGA Tour came to me and they wanted a better putting stat. They moved from putts per round, to putts per green in regulation, to length of all putts, and all had their flaws. They realized that when you rank putters by putts per round, it didn't really match what people thought were the best putters . . . To figure out who's the best putter, you need to adjust for the difficulty of a putt. Because a one-putt from a foot is not a great performance, but a one-putt from 60 feet is a great performance. Even though they both count as one putt on the putts per round scorecard, that's not really a good measure of putting skill."

It took buy-in from the world's best players and coaches to see the new way performance could be measured. It even included a blind surveying of players using a ranking of putters with outdated stats versus a ranking of putters using strokes gained data from the ShotLink lasers. The modern list confirmed the eye test of players as to who were, indeed, the best putters on the PGA Tour.

"They got tremendous buy-in because the results made sense in a way that the traditional putts per round did not," Broadie said. "There was pretty quick buy-in because it was just a more accurate reflection of what everybody knew was going on."

From there, the data was collected over the course of a season from all shots of every player. The ShotLink system enabled tour officials and Broadie to create the baselines from which all performance could be measured. Working backwards from strokes gained putting allowed the areas of off the tee, approach, and around the green to be measured as well, just like in the example of the scratch player before.

The distance variable and strokes gained stats provide more than six hundred different measurements and rankings for the PGA Tour today, a monumental leap from barely a dozen traditional stats used before. Shots from any distance range, any lie (fairway, rough, bunker), any angle (left misses, right misses) are now available to be tracked. With the triangulation of devices mapping greens, tracking the miss tendencies on putts will soon be a large sample

PUTTING PROBABILITIES				
Distance (feet)	One-putt %	Two-putt	Three-putt +	Expected putts
1'	100%	0%	0%	1.001
3'	96%	4%	0%	1.046
5'	76%	24%	0%	1.245
7'	56%	43%	0%	1.440
10'	38%	61%	1%	1.625
12'	31%	68%	1%	1.701
15'	23%	76%	1%	1.784
20'	15%	83%	2%	1.874
25'	10%	87%	3%	1.931
30'	7%	88%	5%	1.977
40'	4%	86%	10%	2.058
50'	3%	81%	16%	2.138
60'	2%	75%	23%	2.214

Using ShotLink data, the PGA Tour was able to show the percentages of putts made by distance and how many putts it took, on average, for a player to hole out from a certain distance. This baseline allowed Mark Broadie and others to create the entire strokes gained model for tour players. *Via PGATour.com*

size as well. With a full generation of golfers now immersed in the data, the game has seismically shifted.

"One thing that I wish that I had was the advanced analytics," Arron Oberholser says reflecting on a playing career that ended right as tour data started to be collected and analyzed. "I think I would've used them a decent amount, but I definitely would've used them for saying, 'Okay, this golf course requires this. You really need to go practice this right here, because this is one of the weak parts of your game.'"

The issue facing statistical analysis of modern performance is that there isn't strokes gained data on Arnold Palmer, or Bobby Jones, or even some of the greats of the 1980s and '90s, whose later years overlapped into the era of ShotLink. Not having baselines to how great players of the past accomplished great feats leaves some guesswork when it comes to if modern analytics has optimized scoring potential simply for a modern game or shown a blueprint that was there all along.

When it comes to the best players in the world, what has been learned from nearly two decades of data? The first is that distance truly matters, more than ever before. The efforts documented in the biomechanics and fitness chapters are fueled by the data science showing that distance has more value than previously thought. Why? The data shows that the gains of hitting it longer than average outnumber the potential losses of a slight drop in accuracy. It is strategically the right move to be more aggressive.

The most visible example of this increased knowledge is Bryson DeChambeau, who famously (and very publicly) bulked up with muscle and speed to take advantage of the gains created by analytical discovery. In the 2020–21 PGA Tour season, DeChambeau led the PGA Tour in distance, averaging 323.7 yards per tee shot (on Par 4s and Par 5s). He ranked 178th (out of 196) in driving accuracy, hitting 54 percent of his fairways. His strokes gained off the tee for the season was +1.162, meaning he was over a shot better, per round, than the average PGA Tour player when it came to hitting tee shots.

Compare that with Bryson's previous best season on tour, 2017-18, where he ranked 25th in distance (305.7 yards) but hit 62 percent of his fairways. That averages out to one more fairway hit per round. That season, his strokes gained off the tee number was +.586.

DeChambeau got longer and sacrificed accuracy in the process but was nearly six-tenths of a shot better off the tee. Why?

Working backwards, the data proved something that was always anecdotally known—the closer you are to the hole, the easier it is to get in the hole with the fewest amount of strokes. But the real analytical math here was understanding how little overall risk was being assumed by the player to guarantee those shorter approach shots. Compared to the average player on tour in driving accuracy, like Bryson was in 2017–18, he is only missing one more fairway now with a more aggressive approach to distance off the tee.

"Missing a fairway does hurt a considerable amount, but he's gained [1.2 strokes] from his extra distance," Broadie breaks down. [In this quote, he is referring to Rory McIlroy in 2018 who had similar numbers off the tee, so the math is the same for DeChambeau.] "He loses 0.3 [strokes, on average] for that extra missed fairway, he's still up almost one stroke per round, or four strokes per four-round event. And that's a big number and it's consistent. It repeats round after round and event after event."

"Every 10 yards you get longer, you save a third of a shot per round," Ben Crane adds. As somebody who turned professional in 1999, he's had to learn so much more about the game he grew up playing without all of the information he has today at his fingertips. "Let's say I'm a 280 guy [280 yards off the tee], and someone's a 310 guy; when we get to the first tee, he's one shot ahead of me. One shot!"

There is no doubt that an improvement in technology has allowed players to achieve bigger distance gains with limited accuracy loss compared to older, inferior technology, and there is no definitive way to say that hitting it further would have guaranteed the same results in a previous era for elite players. It's also difficult, using Crane's example, to find the perfect sweet spot where a longer player seeks more accuracy compared to a shorter player chasing distance. It can be very subjective, so should that philosophy be applied less broadly when it comes to the average amateur player who doesn't have the skill of a professional?

"The distance thing is a double-edged sword," Sanders contends. "Bombers succeed when they have a good week at it and they avoid errors. How much closer does the average player (a handicap of around 15) have to get in order to approximate the same amount of success from the rough as from the fairway?"

Sanders factors that distance to be 75 yards, but the key principle there is errors, all of which can vary in terms of severity. The penalty of potential errors and a low-skilled golfer's ability to hit from the rough can dictate decision-making for each shot and an overall philosophical decision to try and add more distance to one's game. The margin is, understandably, greater for a 15 handicapper than a touring professional. Out-of-bounds or penalty areas skew the potential strokes lost of a bad swing so greatly that valuing accuracy has its place, especially for high-handicap players who struggle with penalty shots and blow-up holes driving scores up. But, is there ever a guarantee that you will hit the fairway?

"If you can't keep the ball in play, that is problem number one, no matter how far you hit," Luke Kerr-Dineen states. "Above all else, you keep the ball in play and then you add as much distance as possible until you can't keep it in play anymore. But the question is, where is that line? And it's different for everybody, but you want to walk right up to that line, I think, as much as possible."

The counterargument to being conservative is that missing fairways with shorter clubs (still likely to happen) has close enough of a miss percentage to the longer clubs to make that decision not as valuable (or risk averse) as a golfer would believe. This is where course and hole strategy have grown into big business, understanding not just when to pick your spots, but assessing risk and selecting aim.

"If there's enough room that I'm going to find my ball between penalty hazards, it's very rare I shouldn't be hitting driver," says Scott Fawcett, who works with countless players through his strategy program, DECADE, which he developed as a competitive player learning from the strokes gained lessons of Broadie.

Fawcett, like many who have immersed themselves in the data to better understand what works best to achieve the lowest scores, looks at both trends overall and a player's tendencies to map out a detailed plan for every round. Gone are the days of finding the fattest part of the fairway and simply hitting at the middle of every green. A second takeaway from the added data, beyond the benefits of distance, is that the knowledge of where players are hitting shots on average has shifted the traditional, one-size-fits-all strategy of the sport.

"The traditional course management lesson, it's kind of middle of the green here or at the pin here," Fawcett adds. "Without knowing what the surrounding hazards are, what the size of the green is, how long the shot is, the player is largely just left to figure it out on their own . . . I 100-percent believe that's from a lack of expectation management, a lack of discipline and patience with where you're actually going to get your scoring opportunities from."

The average PGA Tour player hits approach shots from 100 to 125 yards to 19 feet. That's a wedge for the world's best, and it barely gets to a distance where he will have a 15 percent make rate on the putt. For a 20 handicapper, that average proximity is over 70 feet.

Data tracking companies like Arccos, which has tracked over 300 million golf shots from golfers of all levels using GPS technology worn by the golfer and on the clubs, have worked to manage the expectations of golfers based on their abilities. Their research found that the closer the golfer gets to the hole, the less strokes it will take

From 60 - 80 yards		From 100 - 120 yards	
0-5 hcp	41.38 ft	0-5 hcp	48.82 ft
6-10 hcp	45.20 ft	6-10 hcp	54.56 ft
11-15 hcp	50.49 ft	11-15 hcp	61.85 ft
16-20 hcp	55.21 ft	16-20 hcp	70.76 ft
20+ hcp	62.45 ft	20+ hcp	83.46 ft

Average Distance-To-Pin on Approach Shots

ARCCOS

Collecting data from golfers of all levels has changed the way strategy is approached by players and coaches. *Via Arccos Golf*

to get it in the hole. Laying up to a "perfect number," for amateurs of all levels, does not, statistically, make the next shot easier. Data across the entire spectrum of golfers has changed strategy.

"People are making decisions on their best shots, not on their average shots," says Sal Syed, the founder of Arccos. "What data collection systems like Arccos help you do is make your decisions on your average outcomes because you're basing it on reality. And then your average outcomes are going to improve as a result . . . The perception and reality gap shrinks. For the higher handicap, the more you think you can do, the less you can actually do, and you're making suboptimal decisions."

So, where does a golfer want to aim and prepare for the shot to land?

"In golf, the shot pattern is just so big that it could go anywhere," Fawcett says.

Thanks to launch monitors or in-round data tracking, golfers now have the ability to learn their tendencies. Over the course of hundreds of shots hit, there becomes a circle of misses surrounding a desired target, a probability chart of an area where any given shot with each club is likely to wind up. Fawcett calls it the shotgun pattern. That pattern has misses in all four directions in a 360-degree circle around an average center point. Calibrating that point in the mind and applying it to where the hole is located is the key to that strategy. The hole may be 150 yards away and five yards from the left side of the green, but the center of the player's shotgun pattern needs to be aimed 155 yards away and 12 yards from the left side of the green.

"There needs to be a point in space where half of your shots are left and right of a spot, and that's what I consider your target," he adds. "Really, it should be like 53 percent/47 percent skewed in the direction you're shaping it. So if you're drawing it, your miss needs to be an overdraw. If you're fading, your miss needs to be an overfade."

For the average golfer, the shot pattern should be known and a target line for drives and approach shots fall into a repeatable strategy. The goal, as Sanders's early data suggested, is to avoid the big misses that lead to big numbers. This may mean aiming for the rough knowing one's shot pattern will lead to the fairway a majority of the time.

But what about the world's best players, tasked with finding an edge to give themselves the best chance at a birdie? Where does the strategy to avoid mistakes end and blend with the strategy of trying to get the shortest putt possible? This is where the data understanding is still in its infancy, in many respects. Sound, repeatable strategy wins out over a long sample size of rounds, but professional golfers—especially the elite, major winners—are trying to push the envelope, knowing something special is needed to win.

"You play to win that week, and your approach might change week-to-week," says Arjun Baradwaj, a data analyst for CapTech, which partners with the PGA Tour to analyze and understand its data for both player and public use.

This sentiment of adjusting from round to round is true for any golfer, but is easier to document with data for the world's best. Some weeks, like major championships, par is a good score, meaning the more you can avoid bogeys, the better. Other weeks, simply making pars may not be enough to even make the cut, so finding ways to maximize birdie opportunities is the only strategy forward. There is almost a crossover line in the game where it shifts, and there is plenty of debate about where to draw that line or if it should be drawn at all.

"After you adjust for course difficulty, you get this obvious result that the better players make more birdies and fewer bogeys," Broadie adds. "It's not like one is better or one is worse, or this is the preferred method, or not. [Both strategies] are generating comparable scores on comparable courses in different ways."

But that's for the best players in the world.

"The best players, they make more birdies for sure, but they also have already minimized bogeys to the point where you can only make so few of bogeys," Fawcett says. Once you've gotten to this point, it's kind of impossible to get much lower. But for your traditional player, as you're going from [shooting] 95, by the time you get to 79, you've improved 16 shots, only one of those is from an increase in birdies. The other 15 shots are entirely bogey and higher avoidance. As your scoring average drops from 78 to 75, 82 percent of that is from bogey and higher avoidance. [From] 75 to 72, 68 percent of that is from bogey and higher avoidance. Same thing from 72 to 70, 68 percent of that is from bogey and higher avoidance. So yes, you're making more birdies, but twice as much of the improvement in your score is coming from bogey avoidance versus birdie increase."

It's an important distinction between the average game and the lofty expectations of the professional game. Back to the pros, how much aggressiveness is involved in where players aim and how they set up shots? Sometimes more than the strategy would like or suggest.

"I do look at angles sometimes," says Corey Conners, a PGA Tour winner and one of the best ball strikers in the game today. Conners uses the Shot by Shot system and has been tracking his own stats since he was a junior player. He, like many elite golfers, compare stats and shot tendencies from wins and successful rounds to find patters and strategies that work. "First and foremost, I am trying to hit the fairway, but I am definitely aware of where the hole is located and where you'd ideally want to be coming in to attack."

Is he angle hunting—willing to miss the fairway more often to provide an "easier" approach shot—or simply intelligently aware of proper miss strategy? The same question could apply to his second shot philosophy as well. Answering that question probably depends on the opinion of whether this aim strategy is meant to avoid mistakes or maximize his chance of making birdies. Perhaps the two different strategies are achieving the same thing. Either way, it's a blend of old, pre-data attack philosophy and new, shot pattern understanding of results. And it is still debated.

"We're really asking the question: How much do I not want to be short-sided, versus where's the right place to miss it?" Fawcett concludes. "The right place to miss it is always towards the fat side of the green or on the other side of the green, typically. It's not like you have this magical hourglass-shaped shot pattern where you're either hitting it close or you're missing it in the right place. You have a shotgun pattern, period. And you need to be aiming that shotgun pattern, the length of the radius of it, far enough away from certain trouble, but also getting aggressive enough to actually still have birdie putts."

Some professionals will say that the strategy is thrown out the window when it's birdie-or-bust time. For elite golfers, that works in a short window of time for a round or maybe a tournament with an easier setup. But, calculated planning and strategy is a new tool employed by a majority of golf's best players, and its value is trickling down to all levels.

"Everybody's got a guy or has access to a guy that can tell them, 'Okay, every week, this is what this golf course requires,'" Oberholser adds. "'This is what you need to work on this week to prepare yourself because you don't do this well. And this is what this golf course asks of your game.'"

In 2021, three golfers earned more than 300 world ranking points in the Official World Golf Ranking. Based on those results, one could argue they were the three most successful golfers of the calendar year. All three played full PGA Tour schedules: Collin Morikawa, Jon Rahm, and Viktor Hovland. Their rankings, respectively, in three major strokes gained categories:

Off the Tee—36th, 2nd, 5th
Approach—1st, 8th, 15th
Putting—178th, 42nd, 99th

When Broadie set out to get answers from the data, he wasn't trying to dispute the old adage of "drive for show, putt for dough," but the numbers pointed out some interesting trends, a third takeaway from all of that ShotLink data. Putting had the greatest volatility in terms of results. Bad putters over the course of the season could still have great putting days or putting weeks. Great approach play, however, was a skill that seemed to be more consistent for great players, and a weakness for those who struggled to rise to the top of leaderboards.

"Strokes gained approach, I have found, shows success over the course of a long sample size," Broadie concluded. "Whereas, in some of the best putting rounds, you gain five or six strokes just in a single round. There's really a lot of volatility round to round, and that's partly why the winners tend to be really high up in strokes gained putting for the week."

Couple this with the knowledge of gains from distance and there was a formula at the highest level of golf for players. Hit it far, dial in your irons, and find a way to make putts when it matters most. The percentages dictated that winners would come from that pattern. You can't win without good putting (mostly), but you can give yourself far more opportunities to take advantage of that good putting (outlier?) performance if you can become the best in the ball-striking categories.

What about the flaws in the data? As Broadie alluded to in his comments on bogeys and birdies, course difficulty isn't measured to the extent where it factors in the impact on strokes gained stats. A 100-yard wedge shot with no wind on a soft golf course is judged against the same-length shot on a windy day with firm greens.

When it comes to the all-important distance variable, those proximity measurements don't take into consideration the human element of choice or risk. Missing a green by just 20 feet might lead to a more difficult shot than a 40-foot putt, or in other cases, as Fawcett mentioned, the 20-foot shot off the green may not have been that big of a miss after all. It's where the perspective of the player has to be factored in after the fact, and why many of the world's top players outsource their data analysis to coaches or stats experts to not cloud their minds.

"If I play a round and hit a lot of greens and miss a couple where I am short-sided [not very far from the hole] there are certain times where stats show a deficiency [in the subsequent chip]," Conners adds. He is one of the more hands-on players when it comes to analyzing his own numbers. "The stats show I should hit it close from there, but I can reflect on my game and see if that's really true or if the stats didn't measure the difficulty of the shot properly."

Understanding all of the stats and how it can relate uniquely to the games of average, amateur players is an important part of the learning and discovery process, but committing to getting real data on one's golf game is a surefire way to see advantages and deficiencies.

"You really have to understand and get captured data about your game so that you can pinpoint where you're losing strokes, so you can get better," Syed adds.

"I was much more in the camp [around 2001] of, 'I don't want to know anything. I don't want anyone to mess me up,'" Crane reflects. "Then, Greg Rose [from the Titleist Performance Institute] really kind of changed that for me, by saying the person with the most information, used the right way, wins. It's like, 'Oh gosh. Maybe I shouldn't be afraid of this. Okay, tell me a little more.'"

THE BALL

This ball has had the greatest impact on the game of golf, more than any other piece of equipment in the history of the game as far as I believe.

—Phil Mickelson

The idea of swinging hard isn't a new concept. Every generation of player had a giant who mashed a ball in ways others could only dream. From Ted Ray to John Daly, there have been players with eye-popping power and size that could punish golf balls distances no other player could believe . . . and also directions.

Then, along came Tiger Woods. Bolstered by the first, small wave of technology (more in chapter 7) in clubs and drivers, Woods redefined speed throughout the entire bag. He and his peers started to swing harder. But the harder they swung, the more the ball would spin. It would take off into the air and could change directions like a helicopter. Woods became a marvel because of his complete control.

"Tiger was the first guy I ever saw do with a wound golf ball what everybody is doing now," Arron Oberholser, who grew up playing against Woods, recalls. "He was doing that with a wound golf ball in 1996. You want to know how a guy wins a U.S. Open by 15 shots and wins the Masters by 12? He was already so far ahead of the curve, and then you threw in a golf ball that came along in 2000. But now? Everybody can hit it that way."

That type of ball is the same one that Mickelson is talking about in the quote from 2001 to lead off this chapter. One ball discovery from a team of engineers leveraging new technologies changed the game forever.

"The only radical change in my lifetime was the [Titleist] ProV1 solid core ball," Roberto Castro says. "The golf ball was a very discrete event when it changed."

What does the ball do in the air, or a better question, what is a golfer trying to optimize when a ball launches?

The force that is being transferred from the club to the ball at impact compresses the golf ball in that 0.0005 seconds of inter-action. The ball goes flat on one side, elongates for a fraction of a second as it leaves the clubface and then returns to its normal shape.

The more the ball compresses, the more energy it is absorbing and then using to spring off of the clubface and travel a greater distance. Modern golf balls are built with different compression values to optimize that distance for golfers of varying swing speeds, but the science of compression (while not as varied as today) has always applied to the physics of ball power and flight.

"The compression value of the ball is going to determine how long that ball sticks on the face and how much it compresses before it releases," Chris Voshall of Mizuno says. "Once it releases from compression, then how firm that ball is and how the different layers of that ball interact will determine if that ball is going to launch with more speed or less speed. A higher compression ball is going to release with more speed; a lower compression is going to launch with less speed."

The lower the compression value of a golf ball, the easier it is, naturally, to compress and hit longer distances for those with slow swing speeds. For a golfer with high swing speed, however, over-compressing a golf ball can lead to less control. Higher com-pression golf balls allow faster swing speeds to still compress the ball and get acceptable distance without losing as much control.

"At the end of the day, you want to optimize your ball launch conditions because, after impact, the ball doesn't care what's going on back at the tee," says John McPhee.

After the energy transfer at impact, the construction of the ball, quality of the club, and all of the various launch conditions (discussed in chapter 1) determine the amount of spin applied to the golf ball. When the contact with the club involves any part below the middle equator of the ball (i.e., not a topped shot), you get backspin.

While the loft of the club and its relationship with the angle of attack set the ball on its initial trajectory into the air, it's the backspin that helps it truly fly after launch. The physics term is the *Magnus Effect*, whereby the force of the spin creates more pressure under the ball than on top of it, allowing the ball to stay in the air longer. Eventually, gravity wins, but the proper mix of club (loft with swing creating launch angle), speed, and ball construction is how players dial in exact distances of shots and descending trajectories to hold the ball on the green.

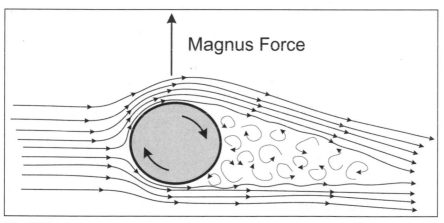

Via Wikipedia

It is also how you combat, or leverage, a ball's left or right movement in the air. Combined with backspin, the sidespin of the ball creates an axis of spin that follows that same effect in the direction of the spin.

The golf ball is also battling the forces of wind and air while it travels through the air. The aerodynamic characteristics of the ball can help it fight that wind resistance. This is why the modern ball has dimples.

As air flows over a ball, there is drag created behind it, an area of wake blocked by the ball traveling through the air. The pressure of air at the front (and top and bottom) of the ball creates this open space behind and creates resistance to forward flight because the object is more attracted to move towards the area of lower pressure behind it. With a smooth ball, this wake is maximized. With an imperfect surface, like dimples, the air flow over the ball is broken up, creating turbulence along the way. This brings the air closer to the ball, minimizing the size of the wake behind the ball and allowing the ball to fly farther through the air.

This is happening with every club. Getting the ball to cooperate in the most efficient way is a major key to the sport and it didn't take long in golf's history to figure that out, even if by chance.

The golf ball is the only piece of equipment that is used for every shot of every round. It is essential to the game and its fine tuning may be more important than any other aspect. Massive advancements in golf ball technology in a short period of time is nothing new in the sport. Today, the ball can be tailored to fit the varying types of swings and speeds. Over a century ago, the ball was engineered more for production and longevity purposes.

While the first golf balls in the recorded history of the sport were (maybe) made from wood, the "modern" game that laid the historical foundation for the sport in the 1800s was playing with a ball known as the feathery. It was a ball consisting of tightly-packed feathers inside a stitched encasement of leather. It was tedious to make, prone to bursting, and expensive by any standards of the time. It's also hard to imagine, without launch monitors two hundred years ago, that the ball transferred energy efficiently or spun in a consistent manner due to the unpredictable structure of its inner and outer makeup.

In the mid-1800s, Dr. Robert Adams invented the gutta-percha ball, or the "gutty" as it would be commonly known. It was made from the hardened sap resin of a tropical Sapodilla evergreen tree. More durable than stitched leather, plus denser than the feathers, the ball was easier to produce, performed slightly better at first, and made golf more attainable for more players.

The original gutty was smooth, but because of its durability, it would get scuffed, scratched, and slightly dented over a long period of use. Consequently, golfers noticed that it performed better than brand-new balls. Through trial and error, golfers discovered the aerodynamic principles by chance, and started creating a more turbulent air flow through the manipulation of the ball's surface. By the late 1800s, every gutty was hammered, scratched, and, eventually, dimpled.

"Ultimately, the golf world chose to absorb the latest technology that society had come up with into itself," says Garrett Morrison, who chronicled the technological advancement of the golf ball—and its philosophical conundrums—through the sport's history in a three-part story series of *The Fried Egg Podcast*. "Gutta-percha, at one point, was the most advanced material in the world, and people were making all sorts of things out of it. And golf brought that into itself. It didn't go any longer than the feathery ball, but it was certainly less expensive. And that was the reason it was appealing."

Enter the dawn of the twentieth century, the Industrial Revolution, and the world suddenly discovered all sorts of new pieces of technology that could be used to improve the golf ball. That meant rubber.

In 1898, Coburn Haskell and Bertram Work, colleagues at B.F. Goodrich Company, invented a new version of the golf ball. Instead of solid gutta-percha, a rubber core was wrapped in rubber thread and then encased with the sap resin of the common gutty ball from the last half-century. A year later, the ball was patented. It created a firestorm.

Suddenly the ball was flying everywhere. What was anecdotally witnessed then is more scientifically known now. First, the improved technology of materials created a significantly better energy transfer from club to ball. Rubber naturally absorbed and moved energy better than the hardened sap of a tree.

Second, the two-layer design of the golf ball was also an engineering marvel. The Haskell ball now spun more efficiently than its predecessors and became the first in a long line of golf balls over the next hundred years that leveraged the benefits of that design.

"The reason why they use multiple layers [today] is when you hit a ball the ball vibrates about its center," McPhee says. "If you think of the different layers in the ball, they all rotate relative to one another like rings."

Put a pin in this idea, as this discovery on how that rotational energy inside a ball controls its flight and spin wouldn't be truly understood for decades, but it's important to point out that the Haskell ball's jump in positive performance wasn't just the exponential gains in distance over previous balls, but also the control golfers gained with the spin being created. The distance, however, was the only thing truly in focus.

The sudden boom in distance changed the way the game was played. While there are no advanced analytics from the early 1900s, it is easily inferred that scoring and play improved for golfers of all levels. Even Bobby Jones famously lobbied for a return to the older ball, opining that the new ball diminished various areas of skill.

The governing bodies of the sport had a decision to make that would set the precedent for how golf ball technology would be approached then and in the future. The USGA established rules that set limits on weight and size in 1921. By 1932, after a period of exploration and discussion, 1.62 ounces was established as the maximum weight, and 1.68 inches was set as the minimum diameter. Those measurements remain the standard today. Notably missing was any rule on material.

"Once the Haskell ball was absorbed into the game, and once the R&A (The Royal and Ancient, the European equivalent of the

USGA) and USGA came to their agreement in the early 1920s about how the ball would be regulated going forward, namely, they would regulate it by size and weight rather than by materials, I think that from that point forward it was a freight train," Morrison says. "I think it's impossible to say whether, at this point, we could take ball technology backwards. I think that golf culture has become accustomed to these advances in technology and that the ball we have now, I just don't see how we're going to change this.

"When something like the Haskell ball is introduced, it really brings into focus I think what people want out of the game. And what most people were saying, especially most Americans at the time were saying was, 'I want anything to help me out.'"

The remainder of the twentieth century brought various improvements on the wound rubber ball first patented by Haskell and Work. The first solution, more obvious and expected in its evolution, was replacing the gutta-percha material as the cover of the ball.

Balata became another naturally sourced rubber-like material that provided a soft, exterior shell. It offered great feel and spin, but also had durability concerns (don't you dare hit it thin!) and, coupled with wound technology, spun the golf ball rapidly for those with speed. There would need to be unnatural solutions to keep up with demand.

"We're now in an era of advanced synthetic plastics, and lo and behold, that's what we have in our golf ball," Morrison adds.

Golf ball manufacturers have tinkered and explored with almost every possible polymer of synthetic material when it comes to making the best golf ball today. While a variety of these materials make up a majority of the balls in production today (Surlyn, Ionomer, Trionomer), the high-end balls used by the best players all feature a urethane composite for the cover. The general difference is feel, cost, and playability, with the non-urethane balls spinning a little less and not wearing as much because the compound is firmer.

Urethane is softer and more pliable for engineers to manipulate, and also "sticks" to the clubface slightly better, allowing golfers to produce more spin on a shot.

The idea of wear, when compared with leather-covered feathers, hardened sap resin, or Balata, is somewhat forgotten today, as plastics have allowed golf balls to be played for dozens, if not hundreds, of rounds of golf for the average consumer.

While each ball designed today has its own proprietary, tested mix of materials in the cover, it's not the only unique design feature of the ball. The USGA does not set rules on dimple patterns or quantity of dimples per ball. Some manufacturers have a more hexagonal dimple, while others are more rounded. Golf balls in the same family of manufacturer can also have upwards of one hundred or so more dimples than a closely related ball. Each setup has a subtle impact on the aerodynamic flight of the ball, altering trajectory and distance based on a golfer's ability or desire. There is no right answer in a sea of balls that are all finely tuned.

"Titleist hasn't had a change in dimple patterns until this year [2021]," McPhee says. "They haven't changed their dimple pattern [for] at least a decade. So, you can't say the aerodynamics have changed significantly. What they have gotten pretty good at is designing the inside of the golf ball with multiple layers."

And this leads to the second, even more radical solution of modern innovation. A wound ball spins, a lot. Replacing that interior with a solid core changed the way it reacted with the club. Various materials would make up the inner parts of the ball through the years—at one point, liquid was a solution—but arriving at a solid core with no material wrapped around it was the first step.

McPhee mentioned the rotation of each layer of the ball around each other. With a wound ball, there is such a difference between the makeup of the solid core and a tensed mass of rubber thread surrounding it that the energy—the speed with which that thread could rotate around that center—was incredible. That meant, however, that the spin happened for every club and there was eventually a moment where there was too much spin, which meant optimized distance was hard to obtain for most players because speed equaled spin, which meant trajectories that were too high.

In the early 1990s, some new balls removed the thread and had a plastic-covered ball with a single, solid core making up the entire inside of the ball, a two-piece technology. The ball went farther, but the playability of shorter shots, which required spin, was not as prevalent. By the end of that decade, however, there was a solution that made its meteoric rise to the bags of almost every elite player.

"There was a diminishing return on swinging faster and that completely went away with the solid core golf ball," Castro says. He was a rising elite junior player at the culmination of the modern golf ball technology revolution. "I think that it's completely changed the game. Of course, [Jack] Nicklaus was the longest. [Greg] Norman, the number one player in the world, had always been the longest, straightest driver. Let's say [length] has always been a huge advantage, but when you could just swing harder and hit it farther with less spin, the game completely changed."

Circling back to the Oberholser story at the beginning of the chapter, all of the sudden golfers could control the spin of the golf ball the way Tiger Woods's game-altering talent could already do with a wound ball. Not only that, you could swing with unapologetic speed. It wasn't just a two-piece (cover and core) solid core that made that possible. While distance gains from a solid core ball made some professionals switch in the 1990s, many held on to the wound ball because of its control on approach shots and around the greens. It wasn't perfect . . . yet.

It took a decade of engineering and tinkering with that one core to realize that multiple solid layers to the core of the ball was the pathway forward. In the fall of 2000, Titleist unveiled the ProV1 golf ball. It was a marriage of both technologies. Solid core for distance, encased in Surlyn, and then wrapped in a urethane, dimpled cover. Now you had three layers to the ball.

"One thing about the modern ball is the multilayers are very carefully set up to give you the spin characteristics that people are looking for," McPhee adds. "If you're clever about your ball design you can design the ball so that you get high spin from a wedge but not as high spin from a driver. Whereas, if you're looking at a single piece or even a two-piece ball, it's very hard to tune the spin rate to give you those different characteristics."

A look inside the modern Titleist ProV1. *Via Titleist Media Center*

It didn't take long to convert the best players in the sport to what the ball could do for them.

"It is the first golf ball that has worked opposite or conversely to every other golf ball that's been made," Mickelson said in a pre-tournament press conference at Bay Hill in 2001, expanding on his quote that led this chapter. "The harder you hit it, the less it spins. Every other golf ball, the harder you hit it, the more it spins. What that does is it gives you a different launch off the driver. It gives you more control of the feel around the greens because the easier shots, where you are not swinging as hard, you have control again . . . It's the best ball that has ever been created . . . It's a ball that fits everybody's style of play."

And that was just the first generation of premium modern golf balls. Other manufacturers started mixing and matching various solid core materials with various covers. Balls marketed to average golfers could now have distance and feel characteristics. Balls marketed for high-handicap players could be both affordable and maximize distance. Compressions could be finely tuned with core materials to make an endless supply of balls for any type of swing and player. The arms race was a blur in the early 2000s that saw, arguably, the greatest advancement of golf technology in the shortest period of time.

"Then you throw in another golf ball even better than [the original ProV1] in 2003, and that's basically where the golf balls stopped," Oberholser says. "For the last seventeen-plus years, the golf ball really hasn't changed much. It's obviously been tweaked and subtly messed with by each manufacturer, but they both pretty much do the same thing from that original ball in 2003, that came out when Ernie [Els] shot 31-under at Kapalua (a scoring record that wasn't broken until 2022) and everybody went, 'Whoa, what's going on here?'"

Cut open any high-performance golf ball now and you might see a rainbow of various colored materials, each of them slightly different in densities and makeup, finely tuned to dance with each other when contacted. It's not just being able to vary speeds and reverse engineer them to different swing speeds, but also engineer the inner materials in the ball to fight the forces that would lead to more sidespin.

The market remains flooded with the right ball for every player. Stronger, variable materials allow engineers to find different ways to tune balls for different swings. There is no longer a uniform ball for the sport, but dozens of different options, making ball choice as important now to the sport as understanding the swing.

"Data should tell you what golf balls you should play with," Sal Syed of Arccos says, reflecting on how little golfers still use growing information in the game to optimize which ball should be played. "Right now, you're just kind of deciding yourself."

On *Golf Digest*'s 2021 Hot List, which ranks every piece of equipment in the sport in terms of fit, value, and performance, there were eighty-five different golf balls evaluated. Eighty-five. Of those, a third were considered winners, maximizing the desired impact on one's game, a far cry from a handful of ball options through the first 150 years of the modern game. An incredible boom for what many consider to now be the most scientifically important development in the sport, with trickle-down effects across the entire golf landscape.

"As the ball has changed, the needs of equipment have changed," Chris Voshall of Mizuno says. "The ball has changed as a result of understanding the true physics of what gets ultimate or optimal. To get the ball that goes the farthest, when the it spun like balata, the ideal ball flight was a ball that took off low and then kind of spun up. Now it's the opposite. It used to be a low-launch, high-spin. Now it's high-launch, low-spin because the dimples have gotten to the point where they can maintain lift at a lower spin rate. I think what we've learned is that optimal has changed along with the golf ball. The golf ball, to me, drives a lot of the changes across the entire bag. If the golf ball still performs like balatas—if non-wound golf balls weren't a thing—equipment would look completely different. But what we've been able to do now is make equipment that speaks to what a ball will deliver."

That begs the question: What's in the bag?

THE CLUBS

If you're a rocket scientist and a physicist, you wouldn't design the drivers we've been playing. That's probably the biggest difference.

—Tiger Woods

Let's first discuss the elephant in the room. All 460 cubic centimeters (cc) of it. No club means more to the modern game than the driver (scientifically, that is; putters are used the most often and, as we learned from analytics, can have the greatest skewing in score). True to its origination as the club meant to propel (drive) the ball the furthest away from the tee and provide (hopefully) the shortest and easiest path to a low score on the hole, modern technology and analytics have made the driver the most marketable, profitable, modified, and perfected club today.

"The advances in driver technology have been really something even over the last ten years," Dr. John McPhee says. "You have this very large oversized head that's very forgiving because it has a high moment of inertia. If you hit the ball off the toe, the driver doesn't twist as much so the ball goes straighter and farther. They tune the location of the center of mass of the driver to maximize the ball speed coming off the face. And the engineering that goes into the face itself is really incredible."

Understanding the forces at play with the driver and how it connects back to the history of the club is important in understanding the technology of today. Unlike the rapidly changing materials of the golf ball through the 1800s and early 1900s, wood was the predominant material used to make drivers until the late 1970s.

Laminated maple and other woods provided a dense material, but persimmon became the go-to wood choice because of its density, while also providing some feel and feedback to the stronger player.

Naturally, it was hard to hollow out wood and lighten the club to help increase speed. Wood-headed drivers were, for the most part, a solid mass, meaning the weight of the driver head was more evenly distributed. That created limits in terms of size because a golfer didn't want the club to end up too heavy, limiting the ability to swing with both speed and control.

That created a small sweet spot on the club—sometimes held together by screws for reinforcement, hence the phrase "on the screws" for a well-struck shot. As clubs would eventually get bigger, the science of the sweet spot is one of the truths that still guides advancement. The sweet spot on the clubface is really an extension of the club's center of gravity (CG), the exact spot in a three-dimensional intersection of lines of the club's mass.

"In terms of CG location there's always been a bit of a debate as to whether the CG should be more forward or back [in the club], but everybody agrees where it should be relative to the center of the clubface," McPhee says. "It should be on a line perpendicular to the clubface through the center of the clubface. And most [modern] drivers have CGs that are within a centimeter of each other."

Everything else, however, was ripe for greater understanding and exploration. The mass of a wood-headed driver was more evenly distributed throughout the club head, creating inefficiency issues with natural materials, especially as engineers started understanding the rotational inertia in the club. MOI (moment of inertia) is the measurement of the club's resistance to twisting. The less the club is able to twist, the easier it is to control face angle at impact, deliver a cleaner strike, and transfer more energy to the ball. To increase a club's MOI, the mass of the club needs to be as much on the outside, away from its center, where the rotational axis twists from.

"What's happening right now is advancements in manufacturing are allowing us to do some things we used to have to compensate for," says Chris Voshall of Mizuno Golf. "To get those things that most players need—the requirements based off of what was manufacturable with what materials were available to process a golf club that launched high and had a high MOI—ultimately the clubs were getting bigger and more offset. The easy way to make a high MOI is to just take a product and make it larger and it becomes more forgiving."

That was the simple solution that started the driver revolution. TaylorMade was the first company to start introducing metal in woods. Metal, lighter than the dense woods being used before, could be constructed in a way where it was more hollowed out and maintained shape, allowing for bigger clubs that offered better energy transfer and forgiveness. By the end of the 1980s, almost every elite player was using a metal driver. Steel gave way to other super metals, until titanium became the metal of choice. Stronger than steel and with less weight, it could be manipulated into making massive club heads, improving MOI, and being combined with other insert materials to dramatically optimize the amount of power delivered to the ball. Higher swing speeds were possible, the new ball wasn't spinning as much, and distance gains were everywhere.

"[Titanium] has variable thickness, so they put more mass in certain spots and they make it thinner in certain other spots to basically act like a big trampoline," McPhee says. "The ball hits it, deflects like a trampoline, and then it sends the ball off with as little loss in energy as possible."

New technology required limits to be set. In 2004, the USGA and R&A jointly adopted new standards for clubhead size and performance. In the simplest terms:

- Clubhead volume will be limited to 460cc
- Clubhead heel-to-toe dimension limited to five (5) inches

- Clubhead sole-to-crown dimensions limited to 2.8 inches
- The coefficient of restitution (COR) of the clubface is 0.830

The first three measurements meant to rein in the almost limitless growth in the size of drivers because titanium and other metals could keep the club head light and strong no matter the size. The fourth measurement dealt with the trampoline effect McPhee mentioned and the growing science that was engineering clubfaces to be as efficient as possible in the energy-transfer process.

COR, in simple terms, is a ratio measurement of energy lost/retained when two objects collide. So, if all of the measured energy from the golf swing that is delivered to the club head transferred perfectly to the ball, the COR would be 1, a perfect transfer. Almost any collision has a natural loss of energy from friction (which creates heat), sound, and other forces in play. The more engineers tinkered with materials, the more likely it was that the COR would push closer to 1. Couple this COR limit with the testing standards of smash factor (see chapter 1) maxing out at 1.5, and drivers (in combination with the modern golf ball) had a maximum threshold to be in compliance with golf's governing bodies.

"Manufacturers are all hitting 0.83 in the middle of the club-face," McPhee adds. "Now the trick is, can you get close to 0.83 for off-center impacts? And that's where that variable thickness face comes in and some companies lately have been using some really sophisticated engineering technologies to optimize that variable thickness of the face."

In the time it took this book to get from writing to press, there has probably been a new development in face technology. In early 2022, TaylorMade launched its new Stealth driver with a proprietary face component called Carbonwood, with carbon fibers threaded throughout the face and bonded with other materials in much the same way golf ball engineers learned how to use urethane and other synthetic materials to make high-performance golf balls. It was a thorough twenty-year research process trying to push the limits of titanium and build on the scientific findings that the lighter you could make the face of the club, the more efficient the impact.

"Because carbon is lighter and less dense than titanium, less resistance and more efficient face flexure lead to a better energy transfer from the clubface to the ball at impact," TaylorMade announced during the launch, in an article written on its website. "The result: faster and more consistent ball speeds."

"The 2022 TaylorMade Stealth with Carbonwood. Another advancement in exploding industry of driver technology." *Via TaylorMade Newsroom*

The company had already created Twist Face technology four years earlier, where the traditional face of the driver was built with over-corrective bends of the face backward on the high-toe side of the face and forward on the low-heel of the face in response to the likelihood of misses from off-center impacts. This was done as an additional advancement to counteract what is known as the Gear Effect.

For off-center hits—especially with larger, modern drivers—the

COR loss isn't the only consequence. Shots that aren't hit perfectly in the middle of the club cause the clubface to rotate depending on the miss. A hit on the toe of the club forces the clubface to open from the force of the impact on that end of the club. Conversely, a strike towards the heel of the face closes the clubface.

The gear effect on the golf ball is what is happening in that 0.0005 seconds when the ball is in contact with the clubface. As the club twists (like a gear) in the direction of the mis-hit, the face is sliding, ever so quickly, across the ball. This imparts additional sidespin on the golf ball because of this movement, meaning toe hits lead to hook spin, and heel hits lead to slice spin. While a majority of the spin on the ball, as discussed in chapter 1, is a direct correlation of club path and face angle at impact, additional (sometimes conflicting) spin can influence that flight based on mishits on the face. A golfer can hit a slice with a 0.0 path-to-face ratio if the ball strikes the heel of the club.

The traditional driver face was never perfectly flat. There was a slight horizontal curve from heel to toe (bulge) and vertically from bottom to top (roll) to help minimize the gear effect of off-center hits. Today, flexing the clubface in different directions is another example of engineering in clubface technology to maximize the driving experience. And much of this advancement is strictly scientific, not even player-driven.

"If you were to put the driver face underneath a microscope, it's thinner in spots that normally would've been thicker closer to the center of the face, and it's thicker towards the outside, in certain spots," says Jonathan Wall, managing equipment editor for Golf. com. "They're able to use artificial intelligence (AI) to design a golf clubface. This kind of feels like an alternate universe, where golf clubs are now being designed by robots and computers and you take the human element out of it. And, look what happens. You get a driver face that looks like that."

This doesn't even begin to cover all of the different advancements in driver heads to make the manipulation of the face easier for golfers. Some weight of the club can be moved by the golfer from toe to heel, or front to back with the twist of a wrench. The face setup can be open, closed, or neutral at address, again done by the golfer with one twist, even between swings (although not in the midst of competition, according to the rules). Once a golfer understands his or her swing path and face angle tendencies, the individual club can be tailored to that swing to minimize mistakes and maximize impact.

"Our job as engineers is to get you to those optimal numbers," Voshall says. "The big trend now is very low spin drivers . . . The spin comes from the impact location in relation to the club's sweet spot and also the given loft of the club. If we can bring the sweet spot lower, then it's going to spin less. And we can bring the sweet spot lower by pulling the center of gravity lower in the head. There's a lot of things that we're looking at to get to these optimal numbers. Those are the levers we're trying to pull."

The engineers aren't stopping anytime soon.

Scientific advancement isn't limited to the head of the driver. The body's choreography to create the power that drives this big chunk of titanium needs a fuel line to deliver that power to the club head. The shaft of the club is vital to the increasing violence of a speed-driven swing, so fitting shaft with swing and club head is a science.

"The shaft, a lot of people don't realize, bends backwards on the back swing, but, just at the impact, the shaft is actually bending forward, which seems a little bit counterintuitive but it occurs naturally from the dynamics of the swing. With that forward bend comes a higher launch angle and a higher angle of attack so it gives you better launch conditions," McPhee says. "Tuning the shaft's stiffness to the individual driver is really important."

In studies conducted by McPhee and his colleagues, sensors on the shaft of the club, coupled with high-speed cameras capturing movement in thousands of frames per second, help to see the shaft's movement, bending, and rotation during the swing. What was discovered is there isn't one point in each shaft that is the main area of flex. The shaft is bending along its entire lens.

"A shaft has a bending stiffness that actually changes along in a continuous way. It's not a discreet thing at all," McPhee found. "They call these curves EI curves, which is the bending stiffness along the shaft. A lot of golf companies are looking into how to come up with the right bending stiffness distribution along the shaft that fits a particular swing style. At the end of the day, you can only have so many products on the shelf, but if there was a way to come up with a handful of different shaft stiffness profiles, and fit them to different swing types, then golfers are going to get better. That's where the research and some of the latest developments are now."

The *EI* term is a combination of two scientific principles: elasticity and inertia. The shaft has an MOI just like the head of the club it is attached to. By applying weights to different areas of the shaft, researchers can test the flexibility of the shaft material (the E) and the thickness of the shaft that would impact the ability of it to twist (the I).

Unlike the club head or the swing, the shaft itself is not creating a significant amount of energy to get to the ball. It is a conduit. The lighter the shaft (modern materials are typically mixes of graphite or carbon fiber), the easier it is to swing faster. The thicker (stiffer) the materials, the harder it is to twist the club rapidly. Intuitively, one would gather that lighter materials would generate more speed and energy for the club head, but a shaft that is too light can be difficult to construct with a stiff enough flexibility for a high-speed golfer, creating a flex (also called deflection) in the shaft that forces the body to compensate, losing energy from the bottom of kinematic sequence—the hands releasing the club head to contact.

There are now aerodynamic limits placed on the shaft by

the USGA and R&A, including restrictions on shafts that aren't smooth (think about the aerodynamic benefits of dimples on a ball; a smoother shaft doesn't cut through wind as easily) and the prevention of any additives that are solely for the improvement of aerodynamics.

Like almost everything else in the modern golf setup, the shaft is tailored to the unique swing, and it can play a large role in the end result of club path and face angle. Get the shaft wrong in a driver and it can become difficult for a golfer to achieve the desired results.

"The biggest purpose of the shaft is to deliver the most energy possible to the ball," Voshall adds. Mizuno was one of the first companies to invest in the shaft-measuring technology of EI. "How is that shaft going to time up with the release of your swing and then also deliver the head consistently in the same orientation from swing to swing?"

Many shafts are marketed with *kick points*, suggesting there is one part where the shaft flexes, but it's really more in reference to where the flexibility (the E) is at its greatest in that continuous bend of the shaft. Fitters combine the knowledge of how a particular shaft bends with the dynamics of an individual swing to marry that unique flex with the way the wrists (discussed in chapter 2) time out the final delivery (release) of power from swing to club.

"In simple terms, you want the shaft bending forwards at impact to give you better launch conditions, to help you get the ball up in the air higher, with less backspin," McPhee concludes. "With a weaker player and a slower swing speed, you need a more flexible shaft so that by the time it gets to the ball, it's still going to bend forward. Whereas if you have [PGA Tour champion and incredibly strong player] Cameron Champ swinging the golf club, he can have a very stiff shaft, and it's still bending forward by the time he makes contact with the ball. And you couldn't give Cameron Champ, or Bryson [DeChambeau], or one of these guys, a flexible shaft. For them, it would be like holding a piece of spaghetti or something.

It would be almost uncontrollable because it would deflect (flex) too much. It's a matter of matching the shaft deflection to the strength of the player."

While most of the science and conversation centers around the importance in the driver because of the speed with which that club is swung, the understanding of shaft technology applies through the entire bag.

"Whether it be a driver or an iron, the goal is the same," Voshall says. "The goal is for the shaft dynamic to marry up to your swing dynamic so that the head is delivered consistently."

And that is where the philosophy of what the driver does shifts to what the irons do.

It has been nearly one hundred years since the USGA and R&A established that fourteen clubs would be the limit one could have in the golf bag. Why that limit was set isn't really relevant to the science of the setup of a perfect set of clubs, but the established limit meant that the gaps need to be filled. Every set will have a putter (which gets its own chapter) and, today especially, a driver (which probably should have been its own chapter too), but every potential shot in between needs a staggered collection of weapons to account for every scenario.

It would be somewhat irresponsible to link all non-driver "woods" into the driver breakdown previously written, but the desired result is largely the same today with the modern golf ball—launch high with lower spin to maximize distance and minimize mistakes. While fairway woods are needed for more playability (most often without the benefit of being on a tee), the science behind their modern construction is still following the same principles of MOI, CG, and COR covered already.

The irons, however, have taken a slower scientific journey to modern optimization and those advancements are still in their early years. For centuries, forged metal, connected on the end of a shaft, was melded and bent into a sporting knife of varying shapes and lofts to rip the ball from the Earth, lift it into the sky, and slice through the turf after impact.

"With an iron, of course, distance isn't as important as consistency, right?" McPhee asks. "Consistent gaps between your irons is very important. You want to dial in a specific distance. With that in mind, with irons, you would like them to be very forgiving so that if you're not hitting it right on the sweet spot, you're still hitting the ball a similar distance."

Irons were solid metal of varying thickness and lacked a ton of forgiveness until around the same time metal woods burst onto the scene. Karsten Solheim and his company, PING, provided the first real leap in technology. By creating an iron that wasn't forged in one solid shape (the common golf term is "blade" or, modernly, "muscle back"), Solheim developed the cavity-back iron, a hollowed-out back with a majority of the club's material around the perimeter, not evenly distributed.

Much like what was learned in driver technology, moving the mass of the club to the outside as much as possible helped significantly with MOI, making the club easier to swing and deliver consistently. Carving some weight out of the back also allowed for larger construction, providing more forgiveness on the face of the irons and a greater area of contact to maximize the club's CG. While there is only one sweet spot on any club, cavity-back irons provided more limited depreciation in distance for off-center hits compared to their blade counterparts.

That discovery would open up thirty years' worth of evolution of cavity-back irons, with sets of clubs built for skill level, many of which offer hybrid setups with shorter, higher-lofted clubs (naturally easier to hit) having a more muscle back setup, and longer, less-lofted clubs (naturally more challenging to hit) having more of a cavity-back setup.

Even with the advancements in forgiveness and the subsequent distance gains of the golf balls and drivers, irons still needed to do their original objective—cover a wide range of distances for approach shots. The onus to maximize the technology (add distance and speed) beyond accuracy took a little while to catch up to the driver revolution.

We are just at the beginning of the "iron speed" revolution, a series of product innovations that are making clubs lighter, stronger, and improve on the COR at impact. Much of this is done with changing centuries of iron manufacturing.

"All the top irons now, they're no longer solid bodies but they're hollow with a very, very thin face and they'll inject some material in behind the face and there's lots of work going on right now to figure out what's the best material to inject behind the face," McPhee says. "It has to support the face. Some of these faces on irons now are so thin that if you didn't put material in behind them, they probably crack without this supporting

Via TaylorMade Newsroom

layer. What you end up with is clubfaces that are hotter (better COR) than they used to be in irons. This is a new breakthrough compared to the solid irons we used to see. The irons are starting to look a little bit more like drivers except, instead of having a hollow space filled with air, they have a hollow space filled with an injectable material."

As McPhee points out, speed is now everywhere in club manufacturing, and it's not necessarily about hitting the ball further, but opening up the ability to use clubs that are easier to swing.

"If you can take the same 7-iron, with the same loft, and you can hit it farther now, that means that maybe you could hit the 8-iron, with a higher loft, the same distance as you used to hit 7-iron," he says. "It gives a little bit more confidence because the higher-lofted clubs are a little bit more forgiving. The speed itself isn't important in order to hit the ball longer, it just means that you hit a higher lofted club the same distance with a little bit more confidence and a little bit less side-to-side dispersion."

It's advanced analytics (shrinking your shotgun pattern) justifying technological advancement. Closer is better, so build stronger irons to make the definition of "closer" an easier club in the bag to hit. The governing bodies took note of this as well, expanding the COR limit of 0.830 to all clubs (except the putter) in 2016.

The relationship between improved iron technology and the new golf ball has seen the control of spin (less than with the wound ball and old iron technology) changing the way irons are built. Because the ball flies through the air better (dimples + solid core), standard lofts on irons have decreased over the years. Some opponents to this development say it's a marketing ploy to sell more clubs that go farther, while club engineers will say that the movement of materials in the irons help launch the ball at the same angles as before, if not even more optimized. Again, it's not a funnel of fitting all golfers into one spec of clubs, it's a system reversing that metaphor of various clubs being fit to one golfer.

"On the iron side, if we can get you more speed, that's great, but from there, we've got to understand, does a player need more launch?" Voshall explains. "The way you get launch is by manipulating the center of gravity deeper. You can get more spin by pulling the center of gravity higher . . . Launch angle and spin rate are going to be determined mostly by the club's center of gravity, where it is located, and the ball speed's is going to be a result of how flexible or how much action you can get out of the face.

"It's just marrying that center of gravity with the dynamics of what a player brings. If a player swings steeper, they're going to impart more spin. If they swing shallower, they're going to impart less spin. If they're more over the top, it's more spin; more from the inside, less spin. It's taking all of these moving parts that a player brings to the table and then marry that equipment to it to get you to an optimal place."

There is a theoretical sweet spot for every club that is unique to every golfer. You don't want every club in the bag, relative to the shot required of it (a driver to go as far as possible; a 9-iron to go an expected distance with a soft landing), to not maximize what it is capable of. There are shots with launches that are too high or too low, or with too much spin or not enough spin, and those shots need to be tinkered to be more efficient.

Back to shafts, do they have an impact on the setup of irons? Certainly, although technological advancements in shafts are seen far more in the driver and wood space. There still has to be an optimal solution to maximize launch and contact conditions, but shaft options are more limited. As the club gets shorter through the bag, its ability to flex diminishes due to simple physics—shorter objects of the same material have less potential to flex, plus shorter clubs have less swing speed potential compared to longer clubs due to the arc and distance traveled. This has prompted less investment in research. Steel shafts are still common on many iron sets, and while its reinforcement, or inevitable replacement, with synthetic fibers or graphite is the current scientific exploration to create better shafts, the tailoring is driven simply by efficiency and speed.

"The shorter [the shaft] gets, you're less reliant on forward bending of the shaft to give you club head speed because you're hitting an iron," McPhee adds. "You're no longer trying to hit up into the ball and have a high impact angle. Now, you're hitting down into the ball. It plays a fundamentally different role than in a driver. I don't think I've ever seen anybody talk about EI and GJ profile (which measures torque) matching for irons. That's really at the driver level. So, at the [average golfer] level, give somebody a shaft that's light enough for them to swing with a decent amount of club head speed."

Iron shafts may be the "simplest" part of the bag to tune, but they add to a growing need to be thorough in finding the right iron combination. To a new golfer picking up the sport for the first time today, gearing up wouldn't be an adjustment because the expectations are all rooted in current technological understanding. To anybody who played golf prior to 2000—even the greatest of all time—the adjustment can take time.

"In order to get the ball in the air, you've got to have spin," Tiger Woods said in a press conference in 2006. "With the opportunity to produce less spin, most amateurs cannot get a 3-iron or 4-iron up in the air . . . We've sacrificed spin for distance, and that's kind of the nature of the game."

Woods's response was to a question about hybrids, a modern technology of (mostly) the twenty-first century that blends wood technology with traditional iron length, loft, and playability. The same benefits in technology dealing with MOI, CG, and COR apply to hybrids and why a 4-hybrid is easier to hit than a 4-iron. It's more options in figuring out which clubs fit swing needs and fill the gaps in that fourteen-club setup.

"Anything that has less spin launches higher and, obviously, it's going to carry further and roll further, and that's kind of where the game is headed," Woods concluded.

More than a decade after that quote, the game has arrived at that point.

When Gene Sarazen hit the "shot heard 'round the world" to win the 1935 Masters, he holed out from 232 yards with a spoon, the nickname for a club with the loft and build that would compare to a modern 4-wood. But it was his club that actually looked and acted like a spoon that created his greatest legacy.

The evolution of the wedge isn't as much technical as it is

serviceable to a golfer's ability to both score and escape. It's hard to imagine it took until the mid 1930s for it to become popularized. (There is some debate as to whether Sarazen officially invented a higher-lofted club that became the sand wedge, but he certainly popularized it in play). There was nothing holding back club makers from bending metal to a more open position to give players the ability to chop balls from thick grass, blast out of bunkers, or hit high and soft shots into greens. A relative lack of green speed (more in chapter 9) likely had some influence on the lack of innovation in clubs lofted higher than a modern pitching wedge (forty-five to forty-eight degrees of loft), but there was no learned benefit yet to what wedges could do for a player.

In the summer of 2006, Phil Mickelson was at the peak of his career. He had won the previous two majors and was poised to win a U.S. Open that had painfully alluded him a title three times before. Winged Foot Golf Club offered a brutal test of skill, highlighted by dense rough that would swallow any ball offline. While Mickelson's eventual second-place finish in that tournament would be remembered most for his errant tee shot on the 72nd hole, it wasn't the driver in his bag that made waves earlier that week.

Mickelson had worked with Callaway Golf to design an almost unheard-of wedge of sixty-four degrees to, as he said before the tournament started, "help get it out of this rough with a lot less bounce, and to help me hit higher, softer bunker shots. Because the bunkers are so deep here and there's so much undulation on the green, I want it coming in as soft as possible."

Finely-tuned wedges have two main principles to aid the golfer in hitting tailored shots: bounce and grind. The bounce of the wedge is determined by measuring the angle between the leading and trailing edge of the club when it is set at a neutral position on the ground. Think of it as the degree of tilt from front to back of the part of the club that will make contact with the golf ball.

The greater the bounce (often ten degrees or more), the higher the leading edge of the club is off the ground at impact. This is useful for players who have a steeper angle of attack or need to play from softer conditions. Having the leading edge higher avoids hitting shots fat.

Conversely, the lesser the bounce (down to four degrees, maybe lower), the lower the leading edge of the club at impact. This is useful for players with shallower attack angles or for shots in firm conditions, where getting the club under the ball perfectly is more difficult because of the ground resistance. In Mickelson's sixty-four-degree, low-bounce wedge, it acted like a razor blade, surgically cutting a path through the rough and sand to get to the ball without losing as much energy. A wedge with more bounce would be more blunt with the leading edge, creating more resistance and difficulty in getting to the ball.

The grind of the wedge is the manipulation or removal of material from the sole of the club to improve the interaction with the ground. It is, simply, the geometric profile of the bottom of the club. There are multiple grinds and almost infinite possibilities of where you could take away material on the bottom of a wedge to alter its shape. Developing grind is as much of an art form and exercise in trial-and-error as it a product of scientific research. With that said, understanding how one's swing tendencies and needs stack up is important to making sure that the wedges in the bag—an extension of the irons in terms of yardage gaps AND clubs needed to hit any number of around-the-green shots—are finely tuned to the golfer for perfect feel and spin.

It all comes back to spin. It was kept low on the driver. It is optimized in the irons. Now, on the wedges, it is essential for controlling how a ball reacts when it hits the green from everything from a 100-yard full swing to a 15-foot chip shot. The modern ball core and advanced plastic cover allows for incredible spin control, but what creates that spin beyond the force of impact? The grooves.

When the ball makes contact with the face of a wedge, it briefly travels up the face of the club in the 0.0005 seconds of impact. That slide creates friction between the two objects and rotates the ball backwards. As previously discussed, that backspin helps keep the ball in the air, but even when gravity brings the ball down, there is still backspin left on the golf ball. How much backspin remains determines which force wins out, the forward momentum of the ball or the back-spinning momentum of its spherical rotation. With drivers optimized for less spin, that means the ball (typically) continues to roll forward when it hits the ground. With a wedge, however, the golfer wants to control that ratio of backspin and forward momentum, perhaps to a point where the ball bounces forward, spins backwards, or stops perfectly in one spot. The engineering of the grooves of the club have improved the management of this for the player.

Grooves serve two major functions. Unlike a smooth surface, they offer a significantly higher potential for friction and more spin. Secondly, like in a tire on a car, the grooves actually channel debris (grass, water, sand) away from the face of the club to try and create the cleanest contact. The thicker the lie (more grass in the rough), the more debris has to be cleared, which is unlikely, leading to shots with less spin and control.

The USGA has added restrictions on groove construction for high-level competition, combatting the rising technology that makes the grooves of the club deeper and sharper to improve all of those characteristics of friction and debris clearance. Whereas Sarazen's original sand wedge was merely an innovative product of loft (and some less-efficient stamped indentions, not modern grooves), the modern wedge has added multiple options to the bag to improve decision-making and shot creation for golfers to have shorter putts.

From there, it's up to the most important club (and maybe skill) in the game.

8

PUTTING

Half of golf is fun; the other half is putting.

—Peter Dobereiner

That famous quote, by one of golf's classic writers, could be tweaked to something even more modern. To many who study the modern game and apply its advanced lessons to the game today, 'half of golf is science; the other half is putting.' No part of the sport is more artistic, unique, and maddening than the simplest stroke per round. It also happens to be the most important when it comes to score.

The formation of the concept of par, no matter the length of the hole, provided an allowance of two strokes per green for putts. Remarkable when you think about the journey to a score. The scratch-golfer standard by which a 400-yard hole is measured would be to guide the ball those entire 400 yards in just two coordinated swings, but then take another two remarkably smaller swings to get the ball from just (say) 40 feet into the hole itself. Golfers today can hit 300-yard drives down the middle of the fairway over half of the time but can't make a nine-foot putt at the same rate.

The average player on the PGA Tour has 29 putts per round. The average high-handicap amateur player has approximately 39 putts per round. A game played through the air (more so today than at any other point in the sport's history) must finish with shots that are mastered exclusively on the ground. There is a science to it—a study that has increasingly grown with the modern game, albeit not with as much fanfare as the technological advances in previous chapters—but it remains a simple formula to get the ball in the hole.

Rolling a ball into a hole is the marriage of line and speed, a mathematical and physics problem which determines the trajectory of a rolling ball set to battle the resisting (or accelerating) forces of friction (from grass) and gravity (from slope) as the ball interacts on the surface of the green. There are better ways to measure all of these today, and more scientific processes (and equipment) are used by golfers of all skill levels to try and master technique. However, when the name of the game slightly shifts from the full-swing goal of getting the ball close to the hole, to the end-game goal of getting the ball into the hole, the stakes feel higher. While every swing is rooted in muscle memory, the number of factors being measured by the brain to hit putts of various lengths and slopes make it, perhaps, the most complicated golfing act when you combine human judgment with human motion. It's a learned skill that is still being learned. To master it, three factors have to come together:

- Reading the putt correctly to determine how the ball will roll and break to get in the hole
- Starting the putt on the intended line to match that read
- Having the proper speed control to manage the ball's break and maximize the chances of going into the hole

This chapter will expand on those factors in reverse order, but before exploring the methods and advancements, it's worth re-exploring how best to measure the success of the end result of that three-step putting progression.

Ben Crane was the best putter on the PGA Tour in 2005 and 2006. Given how new-age analytics would show putting to be the most volatile of all skills, to lead the world's best tour in putting in back-to-back years was a testament to Crane's ability with the flat stick. Strokes gained was just being introduced to the players (and the world), so quantifying what that meant was difficult for Crane, until more than a decade later when he had his statistician go back and crunch the numbers deeper.

"He said, 'What makes you go, Ben, is that you just make an absurd amount of putts from 10 to 15 feet,'" Crane recalls from the education. "So that dictates my practice. I need to make sure that I'm tidy inside of six feet, but I need to make sure that I am exceptional at what I'm best at. It's that mid-range putt that I make at a higher level than most guys. I just thought that was super interesting. I just love it when stats are complicated but help you simplify everything, so that your practice becomes tidy, simple, and very, very clear."

What strokes-gained putting started to show is that the great putters weren't making a lot more putts from long-range distances than others. Being a good putter meant minimizing mistakes in the areas where golfers had the highest concentration of putts per round and maximizing the opportunity to gain strokes with putts that were, effectively, coin flips.

Back to the strokes-gained math lesson . . . The 50-50 make-miss

PUTTING PROBABILITIES				
Distance (feet)	One-putt %	Two-putt	Three-putt +	Expected putts
1'	100%	0%	0%	1.001
3'	96%	4%	0%	1.046
5'	76%	24%	0%	1.245
7'	56%	43%	0%	1.440
10'	38%	61%	1%	1.625
12'	31%	68%	1%	1.701
15'	23%	76%	1%	1.784
20'	15%	83%	2%	1.874
25'	10%	87%	3%	1.931
30'	7%	88%	5%	1.977
40'	4%	86%	10%	2.058
50'	3%	81%	16%	2.138
60'	2%	75%	23%	2.214

Using ShotLink data, the PGA Tour was able to show the percentages of putts made by distance and how many putts it took, on average, for a player to hole out from a certain distance. This baseline allowed Mark Broadie and others to create the entire strokes gained model for tour players. *Via PGATour.com* range for the best golfers in the world is approximately eight feet from the hole. It's not a very long putt. That means that, over the course of a large sample size, PGA Tour players will make half of their putts from eight feet. These putts are obviously varied in terms of slope, speed (uphill or downhill), and emotional importance but, over the course of a season, a player could have 50-plus putts from this length.

In 2006, Ben Crane led the PGA Tour in putting at +0.849, meaning he was that many strokes better than the field each round just with his putter. That season, he made 67 percent of his putts from eight feet, which was sixth best on tour. As for those 10- to 15-footers that he learned about later, he made just over 37 percent, which was fourth best on tour. That's how he arrived at a tour-leading putting season. He was ranked outside the top 100 on tour in makes outside of 25 feet. His putting gains weren't from rolling in bombs (long-range putts).

If it takes, on average, 1.5 shots to hole out from eight feet (a likelier distance after chipping, or maybe validating a superb wedge shot), Crane's strength in this area is gaining a half of a shot on the field 17 percent more often than average. On those mid-range putts he discussed, 10- to 15-footers are holed out only 30 percent of the time, on average in 1.71 strokes, so Crane was gaining +0.71 strokes 7 percent more often than the average player. It adds up.

The flukiness of making long putts underscored something more important when it came to having statistical gains with the putter. PGA Tour players hole 50-foot putts 3 percent of the time, which means it could be months before you make a long bomb. The average hole-out rate is 2.138, meaning it's more likely it will take three putts to get the ball in the hole than one, and that's for the best players in the world! A make would be a gain of +1.138 shots on the field, which is less likely to happen than three-putting and losing -0.862 shots. The best putters two-putt more consistently, which would still gain +0.138 shots on the field.

A part of this success is the ability to make putts inside of six feet, while another component is speed control, which is essential to leaving short second putts. Make rates start to plummet outside of three feet from the hole, so minimizing the amount of three-to-five footers is just as valuable as making them when it comes to strokes-gained math, which is only concerned with how many putts it took to get in the hole from the original distance. This idea is underscored from the highest levels of the game all the way down to high-handicap amateurs. So, is speed control, the third factor to putting, a talent that is measurable?

Justin Silverstein has been the head women's golf coach at the University of Southern California since 2018 and involved with the program long before that. He and his USC team dive deep into strokes-gained numbers on putting. They more than believe in it. They had to track their own numbers, without the benefit of laser-captured data, in both practice and college tournaments. They were patient, gathering dozens of rounds of data and sitting on it before ever trying to generate conclusions. However, the more players practiced, and the more players validated the contact with the ball of their actual putting stroke (or the fit with their putter technology), the less improvement they were seeing in terms of actual strokes-gained putting numbers. It had to be something else that wasn't being tracked.

"We were measuring every other component in the game, but we weren't measuring speed," Silverstein said. "So, we came up with a stat."

For every putt longer than 10 feet, USC players would log the total inches short or long that the putt wound up in relation to the hole. Not diagonally in the cases of putts that missed left or right, but simply the distance past (or short of) the hole the ball had traveled in relation to its theoretical perfect speed of stopping right at (or in) the hole. This would also include an estimate for any putts made, so a putt that went in with a lot of force may have still wound up 40 inches beyond the hole.

They added up the inches, both long and short, over every practice and tournament round, and turned the data over to the statistics department at the school.

"They came out and said that there was a 99 percent correlation between good speed and strokes-gained putting," he adds. Ninety-nine percent!

There was a ratio that the speed data churned out. For every putt left short, that result was catalogued as short inches and the opposite for every putt that went long. They would add up those totals independently and compare the two as a ratio of the short inches and long inches, to see how balanced the speed was over the course of a round. It would also be used in simulated practice drills of 18 varied-length putts in tournament preparation. They charted the likely distance of the first putt on all 18 greens in competition. This, again, led to the realization that there was a balance of shorter putts (the Ben Crane money distance) and long putts where lagging for two-putt results was important for scoring.

What their data showed was that dialing in proper speed was vital to long-term putting success. The coaches had to break many players from the habit of being overly aggressive on putting and put an end to the old adage of "never up, never in." Yes, a putt will not go in if it doesn't have enough speed to get to the hole, but being aggressive with speed on putts outside of 10 feet (leading to an imbalanced ratio of speed control) was more likely to hurt overall putting numbers than help, largely due to the number of longer comeback putts they had, increasing the probabilities of three-putts, which are big strokes losers. A player was best served to hit the putt with a speed that would stop the putt at the hole and play a bit more break. Speed was as essential to overall putting gains in the stats as made putts were.

Dialing in speed control maximizes the opportunity for the

ball to go in. At 4.25 inches in diameter, there is a range of success surrounding the perfect putt where the ball can still get in the hole. While speed control was deemed more important for minimizing the misses, it invariably helped a dedicated golfer find more makes as well. Speed control creates more consistency in the putting performance and also informs the golfer's ability to pick consistent aiming points based on that speed philosophy. Getting it started on the proper line is the second factor, which is where new advancements have really grown to help golfers increase that likelihood over and over.

When it comes to modern technology and its impact on putting, launch monitors and slow-motion cameras have captured data on millions of putts to understand what is optimal when it comes to a ball starting its roll. Unlike full golf shots that need backspin to lift the ball in the air and help it fly, an optimal putt needs topspin applied to the golf ball. Backspin creates more resistance to forward motion on the green and is more likely to redirect a putt offline.

Naturally, putters are the least-lofted club in the bag, but not built to a negative angle because the ball does need a tiny amount of lift at impact. Just like with a full swing, there are optimal launch conditions for a putt, because the ball is slightly depressed into the grass due to its weight and gravity. To get it rolling properly, it must be lifted out of this microscopic depression. That depression has gotten much smaller thanks to modern agronomy (more in the next chapter) helping to create finer putting surfaces with shorter grass heights. Because of that, the standard loft on putters has gradually decreased over the period of 100 years because the ball requires less loft (lift) to get it out of its (now) almost imperceptible depression.

From there, the ball skids before starting its forward rotation. The less skid, the better the chances for a true roll and a good putt. Making sure the putter's capabilities are aligned with the golfer's natural putting stroke helps to minimize this skid and get the ball rolling forward more frequently.

The same measurements applied to measuring driver impact are in play when it comes to putting. A ball starting online is a product of the clubface matching the club path. In the case of putting, that path-to-face ratio is ideally zero degrees, meaning no sidespin is applied to the golf ball, which would push it offline. There is also a smash factor relationship with the putter, meaning that consistent energy is being transferred from the putter face to the ball. Can that smash factor, the COR of the putter, be maximized for off-center hits? The goal isn't more speed (like with drivers), but rather giving the golfer the consistent speed discussed before, enabling the muscle-memorized swing for each distance of putt to be easily repeated and deliver approximately the same amount of force to the ball. Enter putter technology.

In 1911, the concept of a mallet-headed putter was deemed non-conforming by the USGA. Classic putters were typically small blades of flat metal, resembling a butter knife with a wooden shaft extending from the heel. Those restrictions were relaxed over time, but the blade-style putter was predominantly used through much of the twentieth century.

Enter Karsten Solheim again. In 1966, he developed the Anser putter and, like his irons, it changed the way putters were developed.

"That's the original OG, the PING Anser," Jonathan Wall of Golf. com says. "That's been a very popular design that's been copied by other putter brands. When one figures it out, others start to follow suit."

The physics and science of the Anser are the same as the benefits of the designs for other clubs. Part of the back of the club was carved out, moving the mass of the club wider and the center of gravity lower. This, naturally, improves the MOI of the club, giving

The original PING Anser. *Via PING Golf*

golfers more control during the gentle stroke to maintain a square face-to-path angle at impact. He also introduced the offset hosel (where the shaft meets the club) to provide better visibility of the golf ball for the golfer.

Today, the lessons learned from those technological advancements have seen putters grow in all shapes and sizes, looking to find that perfect control of MOI and COR to give golfers a wider-range of potential mishits and more optimized performance. Face technology in putters continues to advance as well, with studies suggesting that off-center hits on putters can create the same gear effect issues as discussed with driver technology. Putts can start (or be spun) inches offline simply by not hitting the CG sweet spot of the putter face. Making the putter bigger and more forgiving is a part of its technological evolution.

"Look at all of the different innovations we've seen just in the shortest club we have in our bag, our putter," Tiger Woods said in 2006. "We can actually produce more of an efficient roll by going through something that's not a classical shape."

Via PING Golf

But putting is still the most subjective club marriage in the sport. While there is data and a science to finding the perfect putter, it's often feel that influences final decision. Unlike the driver and irons, there is almost no limit on how old a putter can be to still be considered optimal for a player. You can hit a 40-year-old driver on the sweet spot every time and it won't go as far as a properly-fitted modern driver with the same exact swing and strike. You can hit a perfect putt with any putter from any era and, at the end of the day, if the ball goes in the hole, that particular stroke is optimized.

In many ways, it comes back to the mental science of the game. With so many putts—often the most consequential shots of a round—being rooted in memory, the putter has the largest emotional connection to the player. That emotion can swing both ways, but it certainly influences outcome. Full golf swings are more athletic. They can be finely tuned. A golfer is in control of every putt and, perhaps, having the perfect partner to dance with on each putt can mean that the connection can't be based solely on data and launch monitors.

"Maybe your putter is the right putter, but we need to cut it down, put a new grip on it, change the loft and line, maybe change the weight a little bit," says Nick Sherburne of club-fitting company Club Champion. "It's not like you have to buy all new. It's just about finding that roadmap, understanding what makes sense, what doesn't make sense; how big of an impact it's going to make on your game."

The stats tell us that strengthening putting from short distances is important and we know that dialing in speed is an important characteristic of putting improvement, but what about reading the greens themselves? This is the final of three factors, and first in the process of the putting routine. Much of it has to do with understanding the topography and agronomy of the greens (much more on grass in the next chapter), but massive improvements in green-reading technology and technique has developed in the last twenty years.

Many golf courses (all at the highest levels of tournament play) have been mapped by 3-D laser cameras, providing full topographical breakdowns of every green surface on the property. This gives players an idea of the percentage of grade of the slope the ball will be traveling on during its entire path to the hole.

Knowing this information on a putt can give the golfer an even better understanding of where the ball needs to start, and how much speed it needs to carry, in order to travel on the correct line to the hole. This added information has provided some stronger mathematical understanding of putting as more and more data has been collected from all golfers.

Knowing the exact green speed is an important number in the equation. Green speed is measured with an older, industry standard piece of technology called the Stimpmeter. Developed by golfer Edward Stimpson in 1935, a Stimpmeter is a simple device that rolls a golf ball from 30 inches up a ramp set to a twenty-degree angle from the green's surface. It is meant to test the speed on a flat surface area of the green and yields a measurement, in feet, of the total distance traveled by that ball. A stimp reading today of eight feet is considered slow by the USGA, while anything 12 and above would be extremely fast greens.

Combining the green speed stimp number with the slope percentage can then provide a more reasonable guess of what line needs to be selected to match those two factors together as perfectly as possible and have the ball roll to the hole in a way that gives it the best chance of going in. Considering downhill and uphill grade is also important when it comes to the speed with which the putt will be hit on that line. Feeling or seeing that uphill or downhill slope adjusts the stimp number, which was calculated on a flat surface. With a hole that is more than four inches wide, there is a margin of error around perfect that still allows putts to be made (think, side door makes), but is there a perfect formula that can remove the guesswork?

Techniques like AimPoint are being used by golfers of all skill levels (including some of the world's best) to measure the degree of slope with one's feet (or read it from the book), and then visualizing an aim point on either side of the hole by closing one eye and holding up the same amount of fingers as slope determined. In an AimPoint blog written by Mark Sweeney in 2017, he simulated a number of putts based on distance, stimp, and slope percentage, with an expectation that the speed used to make the putt would be somewhere between die speed (ball stops rolling at the hole) and a speed that would travel about two-and-a-half feet past the hole. In one example, the research found, a 10-foot putt on a green running eight (8) on the stimp with 1 percent of slope would need the aim point to be 4.1 inches outside of the hole (the high side of the slope/elevation) in order to make. Adjust the speed of the green to 10 on the stimp (faster) and that aim point shifts by 1.7 inches to 5.8 inches outside of the hole. Hit that same putt on a faster green with pace to go by two-and-a-half feet and the point shrinks to just 2.5 inches outside of the hole. The calculations are there, but much of the data is still captured visually, by a human. A 2 percent slope could really be 1.67. Your eyes can be visually confused to seeing slope that isn't there. For even the world's best, there is a big margin of error, which is why that 10-footer is made less than half of the time.

While it's not perfect science (there is a lot of estimating), it is still a very scientific process of understanding how to read greens, which can inform feel. Of course, hours and hours of practice and experience is still the most tried and true method, which is why putting may be the sport's still-perfect blend of feel and technique. As of 2022, full green-mapping books were disqualified from use on the PGA Tour, trying to return some of the skill of figuring out that percentage of slope back to the player's eyes and feel. There is data to show what's working and what's not, but there may not be an exact scientific solution for improving other than finding one's comfort zone moving forward.

"I have this little drill called the coin drill," Crane adds. "You put a coin out three feet in front of you. I learned it from Carl Welty, this coach that I used to work with when I was a kid. You put this coin out there and you roll 10-, 15-, 30-footers over this coin. Can you roll the ball over the coin? If the answer is yes, do you really care where you're aiming? It doesn't matter. One of the three goals in putting is to start the ball online . . . Can you do that? If the answer is yes, game on. Let's now go to more of a competitive practice, where we're trying to make as many putts as we can from six, eight, and ten to twelve feet."

Hard to argue with one of the best putters of the last twenty years.

THE PLAYING FIELD

I've always been a firm believer that golf is two games; it's one on the ground and one in the air. And the more that you bring the ground into play, the tougher it gets.

—Nick Price

In 2013, the USGA was set to hold its national championship at one of the oldest and most storied golf courses in the United States: Merion Golf Club. Located just outside of Philadelphia, the course had more than a century of national championship experience. It also had space constraints to accommodate the modern game.

With a week of heavy, summer rains, there was fear that the course wouldn't provide the ultimate test of golf, which had become synonymous with a United States Open. At less than 7,000 yards in length, Merion was very short by modern golf course standards. Even with the fairways constricted in width to try and increase the challenge, soft conditions from the rain meant players would have no trouble attacking the golf course, right? Hardly.

Justin Rose emerged victorious with a finally winning score of 1-over par, matching the highest score in relation to par for a U.S. Open winner in the last fifteen years. How did it happen? A little experimentation, a little craziness, and a lot of science from Matt Shaffer, Merion's director of golf course operations.

Knowing the best defense against the world's elite players would require firm conditions, Shaffer had to get creative. At one point, he and his staff tinkered with adding subtle amounts of concrete to the sand used to topdress (the process of smoothing, firming, and maintaining healthy grass surfaces) the greens at Merion. Nothing was off the table when it came to firming the conditions.

In the end, Shaffer contracted with a sand supplier who could engineer a sand so fine (85 percent less than a millimeter per granule) that multiple applications of sand in the time leading up to the tournament created greens surfaces at Merion with root structures that were almost impenetrable to water. The grass got a drink and the rest flowed right off. He applied the same philosophy to the fairways, using six hundred tons worth of sand in the leadup. It worked. Balls bounced through fairways and ricocheted off greens without proper spin. The golf course was engineered to be tough because Shaffer, a 40-year veteran of the turfgrass industry, knew his grass.

"When I first got into the business, we had, what I would say is, ten tools in the box," Shaffer said. "We didn't really have sophisticated fertility. Now we have thousands of tools in the box."

"If I had to pick a second [aspect], along with the ProV1 [golf ball], as to how golf has changed in my lifetime, the agronomy is ridiculous," professional golf veteran Roberto Castro says. "Now, it's just incredible, people . . . If you're playing a top 50th percentile golf facility, it's going to be pretty much perfect. And that was not the case twenty years ago."

In 1977 and 1978, the USGA conducted a comprehensive measuring of green speeds at courses around the country. With more than five hundred courses measured with a Stimpmeter, the average green speed was found to be six feet, six inches, the rollout on a flat portion of a green from the end of the stimpmeter itself. For legendary Shinnecock Hills Golf Club in 1978, that number was seven feet, eight inches. Forty years later, when Shinnecock hosted its fifth U.S. Open, the speed of the greens was (officially) listed at 11 feet, although many players contended it was far faster than that. Green speed had gotten 43 percent faster in just forty years. How?

Beyond the obvious technological advancements of grass mowers—the first mowers designed to cut grass at extremely short heights appeared about a century ago and have improved every year since—the science of turfgrass management is big business in golf and any sport that is played on a natural surface. The advancement of that science is a combination of trial-and-error studying within a loyal network of superintendents and the introduction of some very important technology.

"The knowledge these people have on how to manage these greens—and I'm picking greens because they're the highest maintenance— is just astounding," says Dr. Karl Danneberger, a professor of horticulture and crop science at The Ohio State University, one of the leading programs when it comes to educating future turfgrass managers. "Everything moves around; speed, firmness, all this kind of stuff. And those are all based on quantitative types of measurements. Guys are setting up their courses based on these data sets they collect."

What is the data beyond speed of greens that is being captured?

Superintendents are measuring the clipper yield (length and quantity) of grass from every mow to judge growth strength and turf density. They are determining the firmness of greens and turf with a metal ball dropped from a uniform height and then measuring the depth of indentation that metal ball leaves behind. They are measuring the amount of water in the green complexes themselves as a volumetric percentage of the soil's content. The latter measurement, moisture, has been a major key.

"I don't think anybody would deny the fact that the moisture meter has been the most accurate and probably the one [invention] that has really pinpointed real surfaces," says Thomas Bastis, a superintendent by trade who now serves as the PGA Tour's competition agronomist, working with championship courses around the world to maximize playing conditions for the world's best players during tournament weeks. Is there a uniform moisture reading that travels from course to course? No, but the reliability of the technology has allowed professionals to build maintenance plans that meet the needs of each unique course.

"When we look at data for a golf course, it's much like getting vitals, as a doctor," Bastis adds. "Every patient is different. Every golf course is different. If you're trying to compare a golf course to another golf course and then to another golf course, you can find yourself going down the wrong path. When we start evaluating or gathering data, we're really trying to come up with the EKG, the vitals, for that specific golf course that make it tick. Then when we integrate that into what we look at from an agronomic standpoint. We're looking to reproduce a similar product with success. If we have success, we want to mimic that year after year after year, because we have a little bit more of a pathway. We have an insight into what makes that golf course tick at that particular time."

Art has given way to science in many of the ways turf management is conducted. Before, the best superintendents had a feel for what water was needed. Greens, the most vulnerable grass to maintain, were often watered by hand with a hose and a watchful eye. There wasn't immediate feedback on if there was too much or too little water in the ground to sustain life in the grass and the ideal firmness of putting surface. Oftentimes, a superintendent would have to wait hours or days to understand what effect that watering had.

Today, moisture meters allow professionals to get real-time data on the water saturation under the surface. The ability to get those readings at various spots on the green or various areas of the course (especially valuable for courses with elevation where drainage rates differ) allow the superintendents to control the watering, sometimes down to tiny sections. As each year passes, those moisture measurements can be combined with exterior condition data points (temperature, humidity levels, sun exposure, foot traffic) to create a course's blueprint for how to best manage the turf depending on the variables of the day, week, or month.

"It lets them paint by numbers a little bit better and our greens watering has become more efficient because of it," says Chad Mark, the director of grounds operation at Muirfield Village Golf Club, home to the Memorial Tournament. Mark jokes that he misses the old style of checking moisture by sticking a knife in the ground and figuring it out, but he recognizes the level of sophistication that technology has allowed his golf course to obtain. "It takes a little bit of the art out of it, but it does allow for a better product."

The measurement doesn't stop at the watering. Like at Merion for the 2013 U.S. Open, how much sand is added to the soil (topdressing) has a major influence not just on firmness, but on how the grass grows. Some of the firmest golf courses in the world reside near the coast, like the great links courses in Scotland, world-renowned layouts on the sand belt of Australia, or the classic American standards on Long Island in New York. That soil is, literally, millions of years' worth of natural topdressing, creating the effect Merion had to manufacture for its major championship. The grass had already adapted to what the soil provided. Understanding how variable courses will react to the addition of sand, or the aerification process (the punching of small holes into the grass) of bringing the sand (and air) under the surface, is all part of growing data points.

"Are the sand topdressing, the aeration, the cultural practices more in line and you improved from where you were last year? Those are the data points that you sit there and say, 'Hey look, do we have a problem?' Bastis says. "Were the greens firmer? Were the greens softer? Or did they trend the same? And if they trended the same, are we happy with the same, or do we need to now prove that something's more attainable? How much more is attainable is dependent upon, potentially, budget, more sand, more chemical. Can we schedule a different practice? We can't go into that area without some data points."

Today, there is significantly less guessing and golf courses can take the grass closer to its breaking point to create even more perfect conditions, whether that be for regular or tournament play.

"If you're looking at the science of turf, you're looking at this biological system and what would impact any kind of crop or living organism and how that impacts the growth of this plan," Danneberger adds. "From a management style, which is [now] a lot of the sciences, how do you alleviate that?... Superintendents are not paid to manage turf during optimal conditions, spring and fall, they're paid to manage that grass during stressful times, whether it's winter or summer, and that's what they're judged on."

Much of the advancement in dealing with stressful times is understanding the different types of grasses themselves, putting those grasses in the best position to grow successfully and, most scientifically, engineering the grasses themselves to be more sustainable.

In the same way a football (or soccer, or softball) game can play very differently between grass and artificial turf fields, the difference between golf courses simply because of grass type is a unique challenge rooted in the history game. A golf course in Michigan is played on a completely different surface than a course on a Caribbean island. How do architects and superintendents find the perfect grass for a course?

"In general terms, it's their adaptability," Danneberger says of choosing the right grass. "Bentgrass is a cool season grass. It has an optimum condition for growth and, once you get outside of that optimum, then that plant is under stress. And the same with Bermuda grass. There's a reason Bermuda grass isn't used on golf courses up [north]. Even though they've tried in the past, it gets too cold in the winter and it dies."

Those two grasses are the most commonly found types in North American greens, with a dividing line that resembles the Mason-Dixon leaving those courses in the middle debating which grass provides the best opportunity to grow the strongest greens and avoid stress better than the alternative.

For years, bentgrass was thought to be the better grass for

smoother surfaces (greens and fairways). It had, naturally, a thinner blade of grass, allowing it to grow more densely at lower cut heights. Bermuda was a thicker blade of grass, and harder to cut down to a faster speed while maintaining a smooth surface. The higher the cut, the more it allows the grass to fall to one side, especially as it grows during the day and between mows. It's a living organism and will tilt either from slope or from growing toward the sun. This is where grain comes into play and is a major factor in slower greens that needs to be calculated when judging a putt. It's also a factor that is being engineered out of the industry.

"Everybody wanted bentgrass greens," Danneberger recalls from about forty years ago. "You had all these people coming down from the north. They come play these warm season grasses, and they're saying, 'Man, this is like putting on a brush. I don't want this. I remember bentgrass.' So, there was that push and these ultradwarfs kind of came out of that."

The ultradwarf Bermudagrass was a breeding experiment that created some of the finest putting surfaces now prevalent in the sport today. Bermudagrass grows via a process of vegetative propagation, which means it grows from itself, not from seed. Pioneers of new strains of Bermudagrasses would find areas of turf that fit the properties they were looking for in a stronger grass for greens, such as finer leaf blades or the ability to be cut at lower heights. Those areas would be cut out, bred, and combined with other patches of grass showcasing additional strong properties until an ideal new strand of grass was engineered in sod farms. Today, new varieties of grass continue to make greens smoother and faster, not to mention able to withstand harsher weather realities. The engineering is making the entire turf management process different than before.

"We are definitely the recipients of some amazing genetic breeding," Bastis says. "Not to say that our job is getting any easier, but if you look at what goes into tournament setup now, we have a lot more options than previously. Because we've developed a specific strain of grass that can withstand a lower mowing height or better disease tolerance, I think it's more of a mitigation of risk, more so than a performance factor. And not that we're not getting a performance factor, but I'd say more of mitigating risk. We get a little bit more of a sustainability with maybe less inputs. And what I mean by inputs is fertilizer, fungicides, or maybe the plant is a little bit more resistant to wear traffic. By Sunday of tournament, the grass is less beat up. It's got a higher tolerance or there's more plants per square inch."

This engineering goes beyond the traditional two grasses mentioned above. Poa Annua grasses are highly invasive but have made for great putting surfaces at iconic venues like Pebble Beach Golf Links, opening up the possibility of controlling growth to make firmer greens in more places. The same idea is being applied to fescue grasses, long the standard of classic courses in Great Britain, the engineering is opening up real sustainable opportunities not just in the greens, but in the rough as well.

"Virtually every single project that we're doing from the Mid-Atlantic out through the Midwest is all going into turf type tall fescue," says Jason Straka, a golf course architect and student of sustainability, who works on both new and restorative projects, keeping a close eye on what needs to happen to make golf courses both playable and livable moving forward. "If you would've said [fescue] twenty years ago, it would've been these huge thick wick blades and it wasn't anything that you wanted for a lawn or for a golf course certainly. Nowadays, I would say nine out of ten of our projects are using that type of a grass, which have now been bred. It's very difficult to tell the difference between bluegrass (a typical, lush northern grass for rough) and this turf type tall fescue. The advantage is that it accepts wear better, cart traffic in particular, it's more disease resistant, more drought tolerant. It becomes a wonderful grass and the rough typically can be kept in much better condition . . . It's just this continual advancement of technology and how it applies to golf."

What about grass that doesn't even need traditional water? Enter seashore paspalum, an engineered version of an already versatile warm weather grass that now is resistant to the traditional harmful effects of higher saline contents in ocean water.

"If you look at it globally, you can now build a golf course and use a grass where you could never play golf before," Danneberger remarks. "A few years ago, I was in Egypt and courses that were built before the year 2000 were all Bermudagrass. Courses built after 2000 were all seashore paspalum. I say that because things change and evolve too, even if it's with grasses."

That evolution is as much about survival as it is economical for many courses. A warming planet and more extreme weather events lead to a fluctuation in conditions that turfgrass managers must prepare for. This means wider ranges of temperatures and rainfall from season to season. Some elite facilities have sophisticated systems installed underneath the surface of greens that suck the water (think, an underground vacuum) out of the soil until a desired moisture reading is met, while others have systems that can heat or cool the soil temperature to promote better growing conditions. (Augusta National famously touts its SubAir system to regulate moisture and maximize firmness and speed.)

Most courses, however, must deal with Mother Nature's punches more traditionally. With golf courses getting bigger to accommodate the distance gains of players (thanks to the ball, equipment, strength, and analytical understanding), dealing with more turf that needs water and maintenance is a problem.

"When you just look around the country, half the country is under extreme drought heat and the other half of the country is under flood," Danneberger observes. "How do we make managing these grasses in a sustainable fashion? It incorporates a lot of different things. I think that's a big thing to drive what we do and how we maintain golf with as little footprint as possible. Water, both quantity and quality. Second is going to be pest control. It does consume a lot of people's budgets on trying to treat for diseases or insects."

Is the answer more grass? A return to the look and feel of golf courses from the coasts of England and Scotland? With the engineering of stronger grass varieties and hybrids, the answer from an agronomy standpoint is 'yes.'

"There's a very scientific and agronomic aspect to this and that is that grass needs sunlight to grow," Straka says. "Of course, it needs fertilizer and water. When trees are in direct competition with grass for those same resources, trees typically win out. For golf courses where trees are really close to the playing surfaces, it becomes very difficult to manage. And I know some people have said, 'Well, you can artificially take care of that.' Well, what that means is that you have to put in extra amounts of water, fertilizer, and pesticides, because the grass typically will be struggling and not able to resist diseases and insect pests. You have to put in all these extra resources to artificially pump that grass up so that it can compete with trees. From an environmental perspective, you really need to have a balance, and those trees are great, but they need to be kept away from some of the more critical playing surfaces."

All of these factors and advancements will eventually bear more responsibility when it comes to the sustainability of golf courses, but the sudden advancement in turfgrass maintenance and creation has changed the way land is looked upon to host a golf course. There is a strongly subjective opinion as to what an individual wants in his or her golf course, but how that uniquely ideal tapestry is maintained has seen incredible growth, all the way down to the finest grain of sand, which can also be engineered by size and shape to change the feel and playability of shots. Sand that is more angular in construction (many edges) sticks together more and offers easier playability. Rounder sand particles (more spherical) create a softer sand that displaces easier. This sand is more common in older, coastal courses like in Scotland, where the ball plugs more often and clean contact is more difficult. Yes, even the sand can be engineered to meet the needs of the builder.

Which begs the question, what are the needs of the builder? How has the science of understanding better grasses, coupled with the growing statistical data on players of all skill levels dictated how golf courses are designed (or re-designed) to meet the needs of the current sport? Again, a subjective answer depending on what type of challenge the golfer wants in relationship to what the course can provide.

In advance of that 2018 U.S. Open, then CEO of the USGA, Mike Davis, talked about the agronomy of the golf course at Shinnecock Hills. Technology and the engineering and over-seeding of hybrid bentgrasses had allowed Davis and the golf course staff to make the greens and fairways almost any height and speed they wanted. Boundaries could be pushed where grass was cut to never-be-fore-seen heights, but they settled on a more conservative height of four-tenths of an inch.

"We've played U.S. Opens on fairways cut at a quarter inch before," Davis remarked in his press conference. "I think for us, the last several years, we've really come to realize that, at the end of it, this game is about enjoyment. It's about a challenge. I think that over the years what has happened with height of cut, whether that's putting greens or fairways or closely mown [areas around the greens], probably has gotten to the point where it's not necessarily healthy for the game of golf, and we're losing some enjoyment. When I started at the USGA [in 1990], the fairways were, by and large, cut at a half inch or slightly higher. Then we got down to a point where they were being cut down at a quarter of an inch. I realize that's only a quarter inch, but the difference, if you look at a golf ball sitting on a fairway cut at a half inch—and I don't care what the grass is. It could be rye grass, it could be Bermuda, it could be bentgrass—an average golfer can get the ball on the club and can get under it. When you do it at a quarter inch, you have to basically trap the golf ball, and you're basically providing a playing surface that the average golfer simply is not good enough to play.

"We've actually started to raise fairway heights at U.S. Opens. Why do we do that? Well, it's not only better for the average, in this case, members, but I think if [PGA Tour] players are honest too, they don't mind a little more cushion. And by that, you can actually dry the fairways out more so you get a more enjoyable surface to hit off of. It's drier, you get more roll, and it's actually healthier. You lose less water. You lose less nutrients. So, overall, it's good."

He continued, "I would take that same theme to putting greens. There's a bit of a bell curve that's happened over the years in how fast greens have gotten to where they once were, it's hurt pace of play. It's more costly to maintain greens this way. In some cases, it's compromised the architectural integrity of greens. I think for us, a message is a little bit higher height isn't the worst thing both on greens and in fairways. Our message is slow down the greens some, raise the fairway heights, and the game will be more enjoyable for everyone."

The science has become so good that pushing grass to the limit may be something only needed at the highest levels of competition. This leaves all other golf properties the benefit of being able to function within a framework of stronger grasses and sharper monitoring tools to provide even more consistent and sustainable surfaces. From an architectural standpoint, it also allows new courses to be built with a level of sophistication and understanding that past projects would have required years to understand as the course took root.

"A lot of what we do is science-based, but there is also that artistic component to it," Straka adds. Modern architects are more in tune with the soil they are working on, boring underground to get readings and measurements that inform the type of vegetation and build-up of a design. Classic architects were at the mercy of the environment they built on, whereas the first wave of modern architects could move dirt and engineer designs that were sometimes in direct contradiction to the natural landscape. The current approach, with more tools available, is to engineer in collaboration with the environment. Some of the results offer the best of both worlds, a charm of natural landscape with the precision of modern agronomy.

"You still have to spend quite a bit of time organically looking at the golf course and looking at how things are responding to those tools," according to Straka. "Just because we drew it on a piece of paper doesn't mean that's exactly what is going to translate [to the final design]. It's a lot like paint by numbers, versus starting a painting and see how it evolves and massage it from there. That's the art blending with the science."

Straka and his firm, Fry/Straka Global Golf Course Design, presented that modern design philosophy to the world during the 2017 U.S. Open [the course was deigned under Michael Hurdzan/ Dana Fry in association with Ron Whitten and opened in 2006], when Brooks Koepka won his first of back-to-back major titles. He was 16-under par, a 17-shot difference in relation to par to what he would shoot the following year at Shinnecock Hills. Koepka won that 2018 U.S. Open with a final score of 1-over par, the same score of Rose's win at Merion in 2013. How was the same player able to fare so completely differently from a modern venue to a classic one?

"People don't realize that the agronomy effects the scoring tremendously," Castro adds. "Shinnecock is amazing—one of the grand cathedrals of American golf—but it's essentially the same golf course as Erin Hills. It's big, wide fairways. It's big greens. I played both of those U.S. Opens . . . [The scoring difference] is entirely because the greens at Erin Hills were like a video game. They were perfect. It was basically fake turf. You could have a 15-footer and, six inches off the [club] face, you knew whether it was going in. The most perfect greens you've ever seen. Shinnecock's greens are original from 1897. They're perfectly good. They're great. But they're original greens, they're not Erin Hills. That's two, three shots a day. You can get a perfect 10-footer at Shinnecock and it misses because it's bouncing, especially in the afternoon, especially when you're leading the golf tournament and playing late on both Saturday and Sunday. So, to me, Erin Hills and Shinnecock were a very similar test of golf. And to say that Erin Hills was a pushover because Brooks got to 16-under and Shinnecock held its ground because he won at 1-over? I think they were a very similar test and perform very similarly. It was all because of the agronomy on the greens. And that's just my opinion. It'd be hard to quantify that, but I was there."

While that may be over-simplifying (and ignoring) other artistic elements of design (another book unto itself), it's a world-class player's testament to the scientific advancement of turfgrass, which remains the one thing the golf ball (hopefully) comes in contact with the most during a round. Its engineering has made playing surfaces more similarly playable than a generation of grasses ago, not to mention more sustainable for future course designers and players. The ball now travels across modern designs with more control and freedom than courses shaped by an era of limited resources. When it leaves the grass, however, there is no engineering solution to control the next element.

Back to Koepka's back-to-back U.S. Open titles. In 2017, the year he pummeled Erin Hills, the course, softened by rain throughout the week, saw steady, light winds of ten to twelve mph each day (a little gustier in the final round) with temperatures in the eighties. The following year on the New York coastline, rain stayed away during tournament rounds, temperatures were ten degrees cooler, and the winds gusted above twenty mph often through the week. Shinnecock also benefits from the same natural conditions of the sand belt in Australia or the British Isles, a sand-packed soil base that is ideal for creating firmer conditions. Often, the ability to control everything about the golf ball is out of the golfer's hands.

How do nature's elements impact the ball and its travel? Depending on the severity, a lot.

The most impactful is wind, which has two main forces in play to alter a ball's path. The first, and most obvious, is general direction. As the ball loses more and more velocity in the air, the direction of a crosswind will push the ball in that direction the air is flowing. There is no linear calculation that perfectly equates wind speed to distance traveled, as this is influenced by duration in the air, not to mention sidespin.

The best way to learn about the influence of those crosswinds is to learn about the second major influence of wind, which is headwind versus tailwind. Thanks to the knowledge acquired by launch monitors, so much more has been learned about how a ball interacts with wind either in front or behind. One myth that was dispelled using radar technology was that the ball spins more into the wind. It actually doesn't. The other discovery was that, at high wind speeds, a headwind can hurt up to twice as much as a tailwind. How?

Radar findings showed what the eye could see as well. A shot hit into a headwind typically flies higher, travels a shorter a distance, and lands softer with more perceived backspin than a normal shot in controlled (calm) conditions. The reason for this isn't an increased amount of spin on the golf ball, but rather the difference in air speed around the golf ball that creates an imbalance in normalized pressure. In calm conditions, the air hitting the front of the golf ball is traveling at the same speed as the ball itself, creating the drag that allows the ball to lift. Into a headwind, that air hitting the ball is traveling at a speed greater than the ball, which increases the amount of drag. While the evolution of dimple technology has helped to minimize that drag, there is still more of it than with no wind and the gap of air behind the golf ball widens, creating more drag and lifting the ball higher in the air. Upon hitting the green, the ball spins "more" because it is landing at a steeper angle (due to the additional lift) with the usual amount of spin, but it's now able to apply more of that spin than with a standard trajectory.

It doesn't, however, change the traditional teaching that swinging harder will hurt an into-the-wind shot. The harder the swing, the more spin is added, which will lift the ball more no matter the wind, so combining that with the increased drag the headwind imparts on the ball, a harder-swung shot takes off like a rocket, shoots higher into the air, and lands with more backspin ready to rip in reverse.

The impact of headwind creating a shorter shot is, in reality, more about an inefficient ball flight than a ball facing the resisting forces of wind. Hitting more club (less loft) naturally launches the ball at a lower trajectory with less spin, but the lift created by the headwind returns the ball to a flight more fitting of the actual distance and standard club with no wind.

The reverse is true with tailwinds, creating an imbalance in air pressure on the other side of the ball. Unscientifically speaking, there is now a gap of air in front of the ball fighting the traditional forces of drag. The ball has a lack of lift in this scenario. Golfers anecdotally talk about a ball being 'knocked out of the air' with tailwind shots. This is due to the inability of the ball to lift as a result of that imbalanced air flow. Tailwind shots fly with a flatter trajectory, landing with a shallower angle that produces more roll than expected with a normal shot with no wind. The ball certainly goes farther, but by not staying in the air, that distance gained isn't equivalent to the distance lost if the scenario was reversed to a headwind.

The same principles are at play when it comes to both altitude and humidity, both of which affect the density of the air. The higher the elevation, the less dense the air is. The same is true for more humid air because water vapor is lighter at the molecular level than oxygen and nitrogen, so humid air is less dense.

The less dense the air, the less drag can be created on the golf ball. This can impact trajectory in much the same way that tailwind can, so golfers can slightly adjust launch conditions to make sure shots are flying and landing in an optimized way. This knowledge also explains why there was a discrepancy between amateur and professional gains when it came to playing at elevation.

A study of Trackman data found that amateur golfers, at an elevation of five thousand feet, saw an average of six percent gains in distance through every club in the bag. For professionals at the same altitude, the distance gains were above eight percent per club. Why? The professionals were aware that (and capable of) launching the ball higher would combat the lack of lift and drag in less-dense air, optimizing the shot and maximizing the distance gains provided by the air conditions. The amateur, hitting a standard shot, saw gains, but likely less hold on the greens and a lower trajectory than a shot hit under normal conditions.

Which begs the question, is there a standard for normal conditions?

"There was always this big drive for something called normalized data," says Alex Trujillo from FlightScope. "The problem with normalized data is that it uses a model off of what we call scientific sea level, which is fifty-nine degrees Fahrenheit, zero feet of altitude, and 50 percent humidity. The problem with that is, how many people actually play in those conditions on a consistent basis?"

The answer is virtually nobody. From day to day, round to round, the direction of wind or the amount of humidity in the air is changing the very way every shot from every club is played.

As for temperature, colder means shorter, but why? The first goes back to air density. Colder air is denser, so the same principles about elevation are applied in reverse, but an even greater influence deals with the impact and energy transfer of club to ball.

In colder temperatures, the synthetic materials inside of golf balls become more dense and harder to compress. Additionally, the clubface is colder and, with heat being a by-product of the collision between club and ball, more of that heat energy is wasted in a colder environment. All of those factors diminish the energy transfer from club to ball. It helps explain why there is a gradual loss of distance the colder temperatures get, with many studies of data and launch monitors suggesting the loss is around one yard per ten degrees of temperature (Fahrenheit) from an optimal temperature of eighty degrees. Playing a lower compression golf ball than usual in colder conditions can mitigate some of that loss.

Moisture has negative impacts as well. Obviously, playing in rain creates physical resistance in the air, but even if the water isn't falling from the sky, if it's on the ball, it can lead to aerodynamic issues. Water droplets can rest in the dimples of golf balls, and while much of that water will shoot off of a ball at impact, the less depth to the dimples, the rounder the ball acts in the air, which leads to more drag and less optimal flight. Moisture at impact can also cause the ball to skid on the face of the club a bit more than usual, hurting the chances of optimal contact.

All of these weather conditions are measurables, and the data of millions of shots hit in all varieties of conditions allow the most prepared of golfers to understand what will happen and what needs to happen to achieve the most optimized shot.

"We developed these ball cannons where we started firing balls at different weather conditions," Trujillo adds. "We had weather stations plotted along the trajectory, and we came up with this massive calculator, or optimizer, that now gives you the ability to hit a shot in Orlando today, go right into the system, input all the weather for Mexico City, and it will tell me this ball in Mexico City would've done this . . . Now, not only does the player have time to prepare, the caddie can literally sit on the airplane and start preparing for what their shots should start looking like next week when they get there."

Jack Nicklaus gets many people's vote as the greatest golfer of all time. His eighteen major championships is still the gold standard, but his life after playing has a legacy within the game beyond his skill with a club. Somebody who could dominate the courses of his time went on to build the next great golf courses for a new generation to try and dominate. His motivation to do so sums up the joy of the sport within this chapter's lens of grass and weather.

"Why are we building golf courses?" one of his great quotes asks. "Because we enjoy being outside, bringing man and nature together."

While some of nature has been corralled by man, it will never be fully tamed.

THE FUTURE

*Golf is the ultimate decathlon and it is made up of way
more than ten parts.*

—Justin Silverstein

If that quote is as true as it sounds, it's hard to quantify how many new frontiers of discovery there will be in the game of golf as players, coaches, engineers, and stakeholders continue to find ways to make the game better, easier, and more sustainable.

Should it be easier? That's a debate that goes back to the first major leap in golf ball technology one hundred years ago.

"I think that golf is a particularly rich site for that debate because the game is so complex," says Garrett Morrison, reflecting on his own research about the philosophical arguments facing the game during the shift to the Haskell ball in the early twentieth century. "You don't hear quite these same kinds of debates about tennis or basketball. But golf, since we have these varied playing fields and these varied kinds of equipment, I think it serves as a breeding ground for people to have different opinions about what the game should be.

"I think the terms of the debate have changed but today, basically, you still have players who argue that the traditional ways are best, and that rub of the green is a beautiful part of the game, and that we should just accept the mercurial nature of golf as a feature. And then there are others who say that's ridiculous, romantic BS, and let's not deny the advances of technology."

This line in the sand is easily drawn and conflict can be avoided. After all, there is more than enough room on golf courses for both to exist. Or is there? Improvement in all areas of golf science have increased the need for golf courses to get longer. While the average male golfer hits a drive that carries barely over 200 yards, professionals and elite amateurs are optimizing drivers with average distances well beyond 300 yards. There is a trickle-down effect from those gains that creates a real sustainability issue when it comes to land and course construction.

Building a golf course that measures more than 7,000 yards can combat many of those issues, but the reality is that many of the world's courses aren't that long. While only a tiny percentage of all golfers have tapped into the distance gains to make shorter courses obsolete, the perception of reality to the leaders of those courses is that they must get longer to challenge that small subset of elite players and attract new players to the game who are enticed by length and hitting driver unapologetically (remember, analytics).

"If the golf course were at 6,800 yards, it would be fine, but at 6,300 yards, they're having a real difficult time bringing in younger members," reflects architect Jason Straka, who works on renovation projects with many courses that want to get longer. "For younger demographics, some of these courses are becoming obsolete, which is a real shame when you think about it. If they've got the wherewithal, the land, and the money to lengthen it, that's one thing, but some of these places are in core urban areas and they just don't have the ability, or perhaps even the money, to do it."

That doesn't mean that there aren't ways to engineer golf courses to combat the knowledge and technological gains of longer players. The 2013 U.S. Open at Merion is a prime example of that, but it also had almost unlimited resources, time, and required a lot of work to bring the course back to suitable conditions for the average player after being tightened and firmed for the world's best. Agronomically and architecturally, creative minds can do things, but how sustainable are those drastic measures for the sport long term?

"There are other challenges in and around the greens and things that you can do that elite-level players play differently, frankly, than high handicaps," Straka adds. "We do our absolute best to challenge those higher-skilled golfers, while not prohibiting high handicaps from having a lot of fun. And there are ways to do that, whether it's certain types of bunkers, more short grass around greens, different challenges, and we're very, very cognizant of that. You have to be nowadays.

"We already have golf courses that are getting stretched out to 7,900 yards or 8,000 yards. I mean, it is just a physical impossibility. The only way that you are going to take those athletes and those golfers and dial it back down is with the equipment. These guys are smart. They're training harder. It's amazing what they can do physically. And the technology, not just in the clubs and in the balls, but in their training equipment, and their fitness regimen, and their nutrition, is just incredible right now. The only way that you start to temper some of that is to dial back some of the equipment. For that level of golfer, there is no other way."

As early as 1921, there was language used by the USGA in its rules that showed worry about equipment changing the game too dramatically. Limiting power was at the core of many decisions. Three years later, the USGA cited that "the yardage of championship courses has materially increased during the past few years. This is tending more and more to make the championships an endurance test rather than a test of skill. The extra expense in relaying of courses, the purchase of property and the upkeep area [are] also items which convince the Committee that such action is necessary."

That language was repeated often through the next century as many equipment regulations went into place. Limits on driver head MOI in 2006 cited a reduction in "the challenge of the game," and groove restrictions in 2010 cited that "the skill of driving accuracy has become a much less important factor in achieving success while playing golf than it used to be." Through all of this common language, allowances to previous regulations (like how far a golf ball can travel in the air) have been made, bringing into question what the future will allow with equipment. History is murky when it comes to how strict equipment restrictions will be and stay.

For now, the limits to driver head size, MOI, COR, ball size, shaft length (and shape), and a variety of other rules are trying to cap distance to keep courses relevant. What the future holds is largely dependent on if those regulations are ratcheted up tighter. There are scores of engineers and coaches still looking to optimize every golf shot within the framework of the current rules, while still pushing boundaries. New materials in club head construction are inevitable. Perhaps new synthetic materials in golf ball construction will be invented, or more AI development of shapes and solutions in manufacturing will have performance breakthroughs. All of it is fair game as science looks to make the game easier.

"The USGA is getting involved more and more, more engulfed than any other sport," says John McPhee. "If you look at other sports, you don't see the ruling bodies quite so concerned about limiting the engineers and their inventions. And that's probably because [golf's governing bodies] want to limit the length of golf courses. It's also because there's more patents in golf than all other sports combined, probably by a factor of two or three! There's a lot of innovation that takes place in golf that isn't happening in other sports."

Philosophically, innovation may be viewed as good or bad, but pragmatically, innovation is inevitable. Golf is at another interesting time in the sport's history, and while much of the twenty-first century's equipment advancements could be rolled back to limit distance, accuracy, and optimized contact, much of the science is toothpaste that won't fit back into the tube.

While controlling the optimization of equipment is a strong possibility, limiting the optimization of the individual golfer may be impossible. Knowledge is the ultimate power, and with millions of data points informing golfers of all skill levels, human improvement may be as valuable now (and in the future) as the equipment.

"We refuse to guess at anything," says Justin Silverstein, USC women's golf coach. That's the mantra of his team when it comes to improvement and the process. Because of the advancements in sport science, this is how a high-level college program attacks the game and strives to make it better. "I don't have all of the answers, but I do have all of the data," he adds.

The future of that data is what offers a glimpse into new frontiers in the sport. GPS monitoring of golfers from the past decade is starting to influence how architects design new courses or retool current layouts, based on where golfers are likely to hit shots. The same data is already allowing course superintendents to adjust mowing patterns, fertilization schedules, and watering plans, greatly reducing resource use and environmental footprints. That data is still fresh and untouched in many ways.

From an individual standpoint, Arccos golf has seen users of its platform log nearly 200 million shots annually. There is now a baseline for golfers of all levels. That data speeds up improvement, teaches course management, and, maybe most importantly, sets realistic expectations. The learning curve has shrunk and may continue to get smaller as data becomes more accepted.

"There's a huge appetite for data among golfers," says Arccos CEO and Founder Sal Syed. "It makes sense because golf is, I would say, a reflective sport. The National Golf Foundation did a survey for golfers asking, 'Why do you play golf?' and 83 percent of golfers said they play golf to get better at golf. So now, you think about, 'what's the best way to get better?'"

The blueprint to answer that question is currently with the touring professionals, who have embraced the information to drive the sport to record-breaking heights at the highest level.

"The balance has shifted, I think, to a certain extent," says

Arron Oberholser. "There's always going to be technique. There's always going to be room for creativity for artists. But the balance from when I played to now has shifted so much to the data and understanding... It's really, to a certain extent, paint by numbers."

And that's not just the data in shot tracking. It is all of the data that goes into being an athlete. Practice, workouts, course prep, target lines, recovery, club selection, club fit, meals, sleep, all of it, it's part of the sport now, not just figuring it out from shot to shot. It can be overwhelming for an individual golfer to tackle, but the future of the professional game is the strengthening of the current system in place to manage all those areas.

"Better teams; It's more of a group game," says Ben Crane. "Guys have full-time coaches, trainers. Guys are in better shape. They're stronger. They're swinging faster. They're training to swing faster. They're looking at stats. They're attacking strengths and weaknesses ... There's cause and effect. A lot of times you can be working on something that is caused by something that's much deeper. It's so important that your coaching team has a high level of understanding of biomechanics, and club fitting and whatever it is. I think a team that is willing to learn and grow is the best team. If they had written a book five years ago, they'd burn it today, because they've learned so much ... Guys are just flat-out better."

Is there something that isn't measured? What's the next frontier of human improvement?

"The only thing you can't really measure nowadays is the heart of the player," says Mark Immelman.

Is there a way to measure the clutch factor of a player and, if so, how would data scientists even begin to quantify what that factor represents? Tiger Woods won 56 of the 60 tournaments when he led after 54 holes, with one round to play. That astonishing conversion rate of 93 percent is even more outrageous when compared to the

PGA Tour's average rate of conversion, which is approximately 33 percent for all players leading at the same juncture of tournaments.

Was Woods's greatness merely a product of his skill being better than the competition, or was there an additional, inherent ability that allowed him to be mentally stronger in pressure moments? Could he summon perfect shots through sheer will better than others? Anecdotally, the evidence screams "yes," but is there quantifiable evidence? There is hope that shot data could shine a light on what that looks like, through a variety of situations.

"It's considerably more complicated than strokes gained [stats]," says Mark Broadie, who has spent years trying to develop a statistic that measures the "clutch factor" of a player. "There's two components to this. One is, how do you measure performance? And I think strokes gained [stats] are the way to go in terms of measuring performance. But you want to marry that with how much pressure was a player under to get this clutch factor. It's certainly not as straightforward. Say you're within two shots of the lead on the back nine on Sunday. That's clearly a pressure situation, but that's deficient in a number of ways because you could be two strokes behind and there could be ten players between you and the leaders, or you could be five strokes behind and it's just you and one other player. This is putting those two factors together, pressure and performance, and then figuring out a way to rank and communicate who does well under pressure, who wilts under pressure, and in what ways does that manifest itself? Is it in putting, or is it in tee shots or approach shots?"

There is a demand to answer that question even if the data isn't as concrete as putting make percentages. More information impacts future decision-making. It's scientific evolution to try and remove (or at least control) a human element from the sport.

Golf's unlimited tapestry of challenges has inspired engineers to tap into the sport's science and shift the way it can be played by new generations. For a sport that has roots far earlier than almost all modern competitors, however, it continues to thrive thanks to the impossibility of reaching perfection. It is inevitable that more science will be applied to try and create the best version of the sport. While many will try—and some will tap into more of the science than others—the game of golf will remain undefeated.

"I think a lot of the information that has been quantified already existed in the brains of lifers, of experts, of people who spent their whole life caddying or playing," says Roberto Castro. "You've seen more understanding and more analysis, but the net effect is not a different game."

ACKNOWLEDGMENTS

This book was the unforeseen end to a journey that started in 2017 when I launched my podcast, *The Perfect Number*. At that time, I was curious. I wanted to learn more about the sport that had been as much a part of my life as anything else, yet I felt I had no understanding of it whatsoever. For the next four years, a who's-who list of incredibly intelligent guests helped to enlighten my understanding of all of the various factors at play when it came to getting the most out of golf.

To those incredible minds who gave your time to be a guest, thank you. Many of you are a part of this book now because of those chats, which laid the foundation for all ten chapters of exploration.

To a loyal, modest podcast audience of golf lovers who (unknowingly) reached out at just the right moments when I thought the podcast was over, thank you for inspiring me to learn more. Your curiosity is littered through these pages. In fact, the book is for you. When I realized I couldn't come close to writing a book that would further educate the experts quoted within it, I knew who I was writing it for. I hope this different form of media teaches you something and also recalls some of the great interviews that are quoted within it.

Thank you to professional colleague (and *Science of Baseball* author) Will Carroll for connecting the dots to approach me about this project, and thank you to Greg Rakestraw for connecting the final dot. This book isn't written (by me) if not for the belief in the maxim "everything happens for a reason." Will reached out to me the very week I decided, reluctantly, to shut down the podcast that inspired so many of these conversations. It was a sign that this book was the next step, and I am grateful to have been entrusted with the challenge.

Thank you to Julie Ganz, my editor, for that (blind) trust, and for continuing to answer my (probably stupid) questions when I have had them. Admittedly, this step has been far tougher than I could have ever imagined. In my journey to learn new things, I quickly realized how over my skis I could get when consuming the overwhelming amount of material that exists on the world's most complicated sport.

Thank you to those incredible minds who have humbled me in my level of understanding. Folks like Phil Cheetham and John McPhee have dedicated so much time to researching the highest levels of golf movement and technology; to give me some of their time connected so many pieces in this book.

My first interview was always going to be Mark Immelman, one of the kindest and most supportive friends I have in golf, who pointed me in the right direction and whose energy and passion is woven through this book.

To all of the former PGA Tour players who have become (new and old) friends and colleagues—Arron Oberholser, Dennis Paulson, Carl Paulson, Paul Stankowski, Mark Wilson, John Rollins, Mark Carnevale, Roberto Castro—thank you for enriching my understanding of the game through the thousands of hours of experiential learning your careers provided. Whether you were a part of this book or not in word, you are all in here.

To my PGA Tour Radio work family—Fred, Bill, Kevin, Z, Werme, Carn, Doug, Trex, Kelly, Dave—thanks for indulging my nerdiness during so many dinners and days on the road. To Fred Albers specifically, your simple act of recommending me to join this team changed the trajectory of my career and life. Without that, there is no decade on the PGA Tour and certainly no position to learn and be inspired to write this book. I owe you so much.

Thank you to those professional colleagues—Chris Voshall, Mike Carroll, Scott Fawcett, and countless others—who simply enjoy talking about this stuff. You don't have to make time to engage, yet you do, out of passion for the sport and its science.

To Mark Broadie, whose role in revolutionizing the game through statistics inspired my first real step into scientific understanding of the sport while broadcasting it. You have given your time and support on multiple occasions and I am honored.

Finally, to the most wonderful enablers of them all, my family. To my kids, Hudson and Gwen, who think I'm awesome no matter what. To my parents, who never wavered in support of a son who wanted to be a broadcaster when he was thirteen years old, that instillation of chasing your passion helped make this project come to life in many ways.

And to my wife, Mandy, the actual writer in the household, whose simple, one-off line of "Will, you're a good writer" several years ago empowered me more than I can ever express, thank you for giving me latitude and love to dream.

REFERENCES

All PGA Tour statistics referenced in this book are acquired from PGATour.com's public-facing website.

INTRODUCTION
USGA Rules and Interpretations - https://www.usga.org/content /usga/home-page/rules/rules-2019/rules-of-golf/rules-and -interpretations.html#!ruletype=pe&section=rule&rulenum=1.

CHAPTER 1
Real World Physics Problems; Physics of Golf - https://www.real -world-physics-problems.com/physics-of-golf.html.

Lightning Strike—http://www.public.asu.edu/~gbadams/lightning /lightning.html#:~:text=An%20average%20duration%20of%20 time,immediately%20around%20the%20lightning%20strike.

Launch Monitor Data Points (Some clarification on vernacular)— https://mygolfsimulator.com/launch-monitor-data/.

Spin Axis—https://blog.trackmangolf.com/spin-axis/.

Ball Flight Laws—Jeff Mann https://www.perfectgolfswingreview .net/ballflight.htm.

USGA Energy and Force—https://www.usga.org/resources /stemfiles/EF68/energyforce_background_info_MS.pdf.

TrackMan Tour Stats—https://blog.trackmangolf.com/trackman -average-tour-stats/.

TrackMan Averages—https://blog.trackmangolf.com/performance -of-the-average-male-amateur/.

Trevino and Woods chat—https://www.youtube.com/watch? time_continue=123&v=N-MiW_FCORY&feature=emb_title.

CHAPTER 2

"Basic Biomechanics for Golf, Selected Topics"; Phil Cheetham, August 2014.

Science of Golf: The Golf Swing, a video by the USGA—https://www.youtube.com/watch?v=5xVA2MAdimQ.

MacKenzie, S., McCourt, M., & Champoux, L. (2020). "How Amateur Golfers Deliver Energy to the Driver." *International Journal of Golf Science, 8*(1).

Real World Physics Problems; Physics of Golf—https://www.real-world-physics-problems.com/physics-of-golf.html.

Terry Rowles—https://www.youtube.com/watch?v=0E6Beb9Yg6s.

CHAPTER 3

About TPI—https://www.mytpi.com/about.

HIIT Training Benefits—https://journals.plos.org/plosone/article?id=10.1371/journal.pone.0139056.

"The Secret History of Tiger Woods"; Wright Thompson. First published in *ESPN The Magazine* in May 2016.

CHAPTER 4

Morikawa 2021 Open Championship Round 3 Interview—http://www.asapsports.com/show_interview.php?id=167816.

Morikawa 2021 Open Championship Round 4 Interview—http://www.asapsports.com/show_interview.php?id=167871.

"Brain Training Lesson with Debbie Crews"—https://www.youtube.com/watch?v=CkKOZ5qvBBE.

"What Should Be Going on in Your Brain During Golf with Dr Debbie Crews" an article and podcast with Golf Science Lab—https://golfsciencelab.com/golf-state/.

Opti Brain Supporting Research—https://www.myoptibrain.com/research/.

Mickelson PGA Press Conference—http://www.asapsports.com/show_interview.php?id=165288.

Morikawa 2021 Pre-Tournament Ryder Cup Press Conference—http://www.asapsports.com/show_interview.php?id=169486.

"The Brain Science Behind Golf: Why Some Experience the Yips" a presentation given to The Center for BrainHealth—https://www.youtube.com/watch?v=uHXpFWqCl4U&t=690s.

Whoop Background Information and Resources—https://www.whoop.com/thelocker/sleep-debt-optimal-playbook/.

"Your Brain on Confidence" by Caroline Centeno Milton for *Forbes Magazine*—https://www.forbes.com/sites/carolyncenteno/2018/04/18/your-brain-on-confidence/?sh=801c4b60cb68.

Golf: The Ultimate Mind Game by Rick Sessinghaus. Excerpt from page 22.

"Does Thinking Burn Calories? Here's What the Science Says" by Markham Heid for *Time* magazine September 19, 2018.

Justin Thomas Heart Rate Data—https://golf.com/instruction/fitness/justin-thomas-heart-rate-data/.

Van Dongen HP, Maislin G, Mullington JM, Dinges DF. "The cumulative cost of additional wakefulness: dose-response effects on neurobehavioral functions and sleep physiology from chronic sleep restriction and total sleep deprivation. Sleep." 2003 Mar 15;26(2):117-26. doi: 10.1093/sleep/26.2.117. Erratum in: Sleep. 2004 Jun 15;27(4):600. PMID: 12683469.

CHAPTER 5

PGA Tour ShotLink Explanation and Baselines—https://www.pgatour.com/news/2016/05/31/strokes-gained-defined.html.

Arccos Data for Amateurs—https://www.arccosgolf.com/blogs/community/course-management-101-what-layup-yardage-is-your-sweet-spot.

World Golf Ranking Points from 2021—http://www.owgr.com/archive/PastRankings/2021/Rankings/owgr52f2021.pdf.

CHAPTER 6

Phil Mickelson quote on the ball from 2001—http://www.asapsports .com/show_interview.php?id=12802.

Physics of Flight in Golf Ball—https://www.youtube.com/watch? v=fcjaxC-e8oY.

Why Golf Balls Have Dimples—https://www.youtube.com/watch? v=fcjaxC-e8oY.

Fried Egg Stories, The Ball—https://podcasts.apple.com/gb/podcast /fried-egg-stories-the-ball-part-1-gutty/id1131723994?i =1000485159795.

"Bobby Jones Wanted to Roll Back Golf" for GolfCompendium .com—https://www.golfcompendium.com/2019/12/bobby-jones -wanted-to-roll-back-golf.html.

History of Equipment Rules, USGA—https://www.usga.org /content/dam/usga/pdf/Equipment/History%20of%20Equipment %20Rules.pdf.

Golf Digest 2021 Hot List—https://www.golfdigest.com/story /golf-ball-hot-list-2021.

CHAPTER 7

Tiger Woods comments on technology from 2006—http://www .asapsports.com/show_interview.php?id=40458.

"Golf Club History: Woods and Irons" by Ryan Barath for GolfWRX— https://www.golfwrx.com/618340/golf-club-history/.

"Physics of the Golf Club" by the USGA—https://www.youtube .com/watch?v=-_WZYKGlDvY.

History of Equipment Rules, USGA—https://www.usga.org /content/dam/usga/pdf/Equipment/History%20of%20 Equipment%20Rules.pdf.

Coefficient of Restitution Definition—https://en.wikipedia.org/wiki /Coefficient_of_restitution.

TaylorMade Carbonwood Launch—https://www.taylormadegolf
.com/clubhouse/244644-a-20-year-journey-to-the-carbonwood
-age.html?lang=en_US.

Gear Effect—What Is It? By Adam Young—https://www
.adamyounggolf.com/gear-effect/.

"Everything You Need to Know About Golf Shafts" by Golf Science
Lab—https://golfsciencelab.com/everything-need-know-golf
-shafts/.

"EI Shaft Profiling" by Russ Ryden—https://www.golfshaftreviews
.info/ei/.

Gene Sarazen wiki—https://en.wikipedia.org/wiki/Gene_Sarazen

Phil Mickelson pre-tournament press conference at 2006
U.S. Open—http://www.asapsports.com/show_interview
.php?id=38016.

Wedge Bounce Explained; Titleist—https://www.vokey.com
/explained/wedge-bounce#.

Friction and Spin, USGA—https://www.youtube.com/watch?
v=TOMIgdxqAEQ.

CHAPTER 8

Putting stats via Arccos—https://www.arccosgolf.com/blogs/
community/tracking-putts-is-the-key-to-lowering-your-scores.

"The Key to Putter Fitting: Know Your Roll," by GolfWRX
—https://www.golfwrx.com/268245/the-key-to-putter-fitting
-know-your-roll/.

The Story of the Anser Putter—https://ping.com/en-us/media
/articles/evergreen/anser-putter-technology-triumphs.

"Robot Data Reveals How a Minor Heel or Toe Miss Can Penalize
Your Putting" by Jonathan Wall—https://golf.com/gear/putters
/robot-data-putter-miss-will-zalatoris/.

The Stimpmeter—https://en.wikipedia.org/wiki/Stimpmeter.

"Picking a Line, What's the Point?" by Mark Sweeney. AimPoint Blog—https://aimpointgolf.wordpress.com/2017/08/20/ picking-a-line-whats-the-point/.

CHAPTER 9

Nick Price quote and various quotes on architecture, 2018 Architecture Forum conducted by the USGA at the US Open— http://www.asapsports.com/show_interview.php?id=140990.

Matt Shaffer, Merion profile article written for the Ohio Turfgrass Foundation Annual Conference in 2014.

2013 U.S. Open Course Setup—https://www.gcsaa.org/uploaded-files/newsroom/tournament-fact-sheets/2013/usga/2013-u-s-open-championship-fact-sheet.pdf.

"A Short History of Green Speeds and . . ." for cureputters. com—https://cureputters.com/blogs/news/a-short-history-of -green-speeds-and-the-evolution-of-the-modern-putter-loft -and-weight.

"The Architectural Speed Limit for Putting Greens" by the USGA—https://www.usga.org/course-care/forethegolfer/2017 /the-architectural-speed-limit-for-putting-greens.html.

2018 U.S. Open Course Setup—https://www.gcsaa.org/docs /default-source/tournament-fact-sheets/pga-tour/2018/us-open -championship-2018.pdf?sfvrsn=9475ee3e_2.

"Headwind vs. Tailwind" by TrackMan—https://blog.trackmangolf .com/headwind-vs-tailwind/.

"Weather or Not: Prepare for all Golf Conditions," by Arccos Golf—https://www.arccosgolf.com/blogs/community /weather-or-not-prepare-for-course-conditions.

Air density explained for humidity—https://www.chicagotribune. com/news/ct-xpm-2011-06-14-ct-wea-0615-asktom-20110614- story.html.

"How Altitude Affects the Distance Your Ball Flies" by TrackMan —https://blog.trackmangolf.com/how-altitude-affects-the -distance-your-ball-flies/.

CHAPTER 10

History of Equipment Rules, USGA—https://www.usga.org /content/dam/usga/pdf/Equipment/History%20of%20 Equipment%20Rules.pdf.

TrackMan Averages—https://blog.trackmangolf.com /performance-of-the-average-male-amateur/.

Arccos Year-End Data—https://www.arccosgolf.com/blogs /community/end-of-year-recap-2021.

"Closing the Deal: Examining the Success Rates of 54-Hole Leaders" by Justin Ray for 15th Club - https://www.15thclub.com/2019/11/13 /closing-deal-examining-success-rates-54-hole-leaders/.